The Backyard CHICKEN BOOK

The Backyard CHICKEN BOOK

A Beginner's Guide

Edited by

H. Lee Schwanz

Skyhorse Publishing

Revised and updated from the book originally published as *The Family Poultry Flock* © 1979 by Farmer's Digest, Inc.

All photos are from Thinkstock.

10 9 8 7 6 5

Library of Congress Cataloging-in-Publication Data is available on file

Print ISBN: 978-1-62914-204-3
Ebook ISBN: 978-1-62914-274-6

Printed in China

This book is dedicated to my mother, Rae Cafferey Schwanz, who taught me most of what I know about chickens. She ran the poultry enterprise on our Iowa farm a generation ago. As soon as I was big enough to carry a bucket of feed or a pail of water, I became her assistant.

She taught me to wonder at the miracle of the hatching chicks and the joy of watching the downy bird learn to scratch for life. She also knew how to fry a chicken the way it could make a small boy's stomach squirm with anticipation.

We hatched about 500 chickens each spring. These were straight-run chicks, so we had plenty of roosters to eat all summer and fall. Each Saturday, year around, there was at least one crate of eggs to haul to town and sometimes more.

Looking back, the chicken business really hasn't changed that much, at least for small flocks. The equipment shown here is virtually the same. The Leghorns we raised still are the most popular egg breed.

Our farm was a leading example of what now is known as the "Protestant work ethic." This, boiled down to its basic meaning, is "If it's hard work it must be right." My father was very progressive in field work and in livestock selection but we really made hard work out of chores. Every pound of feed and gallon of water was carried more than 100 yards both to the hen house and to the growing chickens.

If I could go back over the years and give my mother one gift, it would be running water for the chickens and a more willing boy to help carry the feed and gather the eggs.

Contents

Introduction

OUR FAMILY MADE THE DECISION to be in the egg business and a flock of 400 or so hens was large for that time. Most farms had a few birds of the heavy breeds that got very little care. They scratched around the barnyard picking up a kernel here and a kernel there. Hunting eggs around the barn was an exciting chore for kids.

Our hens were confined in a two-room henhouse. Pullets were in one room and the yearling hens in another. We gathered eggs twice a day, candled them in the basement, and took them to town when the 30-dozen case was filled.

We sold cattle and hogs a half-dozen times a year, but it was the eggs that bought the weekly staples at the grocery store.

Like other farms, we also relied on the flock for food. There were eggs for breakfast every day and two or three for the hard workers in the family seemed about right. We ate a lot of fried chicken, too.

In this chapter, we talk about making a decision on the kind of poultry enterprise you want for your family. Egg income plus family food was our goal. It fit our family, our farm buildings, and the market situation at that particular time. Your needs, facilities, and desires are different . . . select your chickens to fit your own individual family situation.

Chapter 1
What you can expect from a family poultry flock

A SMALL FLOCK OF POULTRY can supply all of the eggs your family needs during the year. There also can be some broilers for barbeque or frying. After the hen has completed her egg-laying days, she can provide the family with chicken stew or other dishes.

A small flock takes a relatively small amount of space. Management is not difficult, but the flock requires attention every day. Someone in the family must care enough about the flock to make sure they have feed, water, and egg gathering on a regular basis.

The first step in deciding on a family flock is to determine what you really want. You can go with a laying breed such as the White Leghorn that produces top quality eggs but provides little meat. You can select a dual-purpose Rhode Island Red or New Hampshire that provides fewer eggs but much better meat. Another alternative is to select fast-growing broiler chickens that are ready to eat in eight weeks or less.

Whatever type you select for the home flock you must realize that you probably aren't going to make money if you put a value on your time.

Commercial egg production is a big business. These farms are mechanized and highly efficient. They can put eggs in your supermarket for little more money than it will cost you to produce them.

The broiler business also is concentrated in giant factories. Chicks are raised on computerized feed formulas and are ready for market in six weeks or less. Automated packing plants process the birds, package them, and deliver them to stores in big volume. Again, it is tough to compete.

Set your sights on a flock that provides for the needs of your own family. Take your reward in fresh eggs and meat. If you can sell the surplus, consider that an extra benefit.

How big a flock?

Determine how many hens you will need to supply your daily consumption of eggs. It may be a month from the time you find the first egg until the hens are laying at their highest rate. Twelve hens at their peak of production will probably produce a total of nine to ten eggs daily for several weeks—perhaps for two to three months. As they age, they will gradually lay at a lesser rate until the twelfth month when you could expect six eggs each day. Don't be surprised at daily fluctuations. On some days, all twelve hens may lay; on other days, only seven or eight. Hens are adversely affected by bad weather, dark days, severe cold, frozen water, or lack of feed.

Most family laying flocks are in the 20 to 25 hen range. It doesn't take any more time or effort for a few extra birds.

The average hen usually produces eggs economically for 12 months. During that time, she will lay from 17 to 19 dozen. She will also eat an average of five pounds of feed per dozen eggs produced.

What are the costs?

It is about as difficult to estimate costs for the home chicken flock as it is for lawn and garden projects. People have widely varying attitudes about investing in items that add interest to the project. Logic and strict economy do not typically govern all decisions.

Some persons will have poultry housing for little or no cost. Others will want to use their handyman skills in the home shop to build an attractive unit which will fit well into the landscape.

Ask yourself these questions

Your answers to the following questions will help you decide whether you want and can afford to raise your own poultry.

1. Do zoning laws permit raising poultry at your location?
2. Do you have "unused labor" available?
3. Is someone willing to care for the birds daily?
4. Is someone able and willing to butcher the meat birds at home, or is there a facility nearby where you can have your birds custom processed?
5. Do you presently have the necessary housing and equipment, or will you have additional expenses for these?
6. Money spent for housing and equipment becomes a poor investment unless it is used for several flocks. Do you plan to continue raising home flocks for several years?
7. Can you use your facilities or proposed facilities for some other purpose (such as storage) if you do not continue to raise poultry?
8. Can you reduce the feed costs by using home-grown grains and/or pasture?
9. Are your facilities or proposed facilities designed and located to prevent causing a noise, odor, or fly nuisance for your neighbors or your own family?
10. Do you have a freezer, so you can make best use of the meat birds you grow?
11. Do you have neighbors who would like to buy "home-produced" eggs or poultry when you have more than you can use?

Your answers to these questions will probably point out some disadvantages of growing your own poultry.

On the other hand, some advantages exist which often cannot be given a monetary value. Whether real or imagined, some people feel the home-produced birds and eggs are better. Certainly, they would be fresher. You can grow the meat birds out to the size or sizes you prefer. For example, with chickens, you may want to slaughter part of the flock at 7 to 9 weeks of age for broiler-fryers and keep the remainder for 12 to 15 weeks for roasters.

If a "dual-purpose" breed of chicken is raised, you can slaughter the males as broiler-fryers or roasters and keep the hens for egg production. The hens will provide some baking or stewing chickens when new layers are brought in.

There is also an intrinsic value to having living and growing animals, especially around children. Children can handle many of the day-to-day chores of growing birds. Minimal space and housing are needed for small flocks. The idea of producing something for themselves appeals to many families.

Be a good neighbor

When you have chickens in a suburban area it is very important that you be a good neighbor. If your operation is unsightly or smells, you are going to have problems.

Keep the area around your poultry house attractive, reducing odors and nuisances to a minimum, and establishing an open door policy.

The area around a poultry building is often neglected and frequently becomes untidy and overgrown. Piles of rubbish and weeds can interfere with ventilation and provide a harbor for rodents and flies. All poultrymen should have a landscape plan. It should consider ground cover, prevailing winds, types of trees and shrubs, drainage points, and access roads. The plan may be completed all at one time or in stages.

The area around a building can be left bare, covered with crushed rock, or planted to grass. Although bare earth and crushed rock are the easiest to maintain, a lawn is generally most satisfactory.

Disposal of excess water, especially from waterers, is often a problem. Simply running it out the ends of the building is not satisfactory. Stagnant pools or improper drainage may result. Give careful consideration to the disposal system.

Proper manure management is essential, because most of the odor associated with poultry comes from the manure. The secret is to keep the waste material dry. Spreading is a critical operation, too. Always pick a good day—never weekends or holidays. Avoid hot, muggy days and those times when the wind is blowing toward a neighbor's residence.

Flies generally pose the most serious nuisance. A sound fly-control program should be followed. But remember, any successful program involves more than an occasional chemical treatment or a general clean-up of the premises. It's a job that needs attention every day, just like gathering eggs.

Dead birds must be disposed of promptly. If this is not feasible, put them in a tightly covered container.

Building goodwill

Communications are extremely important. Tell your neighbors and local businessmen what you are doing. Invite them to see your poultry house.

The appearance of a poultry operation can be quite attractive. But when weeds, brush, junk, and obsolete equipment accumulate, the poultry unit gets a bad name. Then, the poultryman has no one to blame but himself when neighbors complain. Your good name in the community is worth a little paint, a few shrubs, and a bit of consideration for the other fellow. Public relations are more than a job. They're an obligation.

Chapter 2
Which breed is best for you?

WHITE LEGHORNS WERE THE BREED selected for my family's egg-oriented chicken enterprise. Even in those days, Leghorns had the jump on other breeds for egg production. Careful selection in the years since has improved average production of the breed by 50 eggs per hen, but Leghorns were the leaders during my youth. Most farms favored the heavier breeds. They made better fryers then and still do. They also were able to forage for themselves better than a Leghorn. Not much has changed over the years except that both types of chickens have been improved. The Leghorns lay better and the dual-purpose breeds turn out more eggs and grow meat faster. It isn't quite true to say that we were exclusive Leghorn growers. My grandparents lived in a separate house on our farm and they raised Rhode Island Reds. These were heavy breed chickens that laid brown eggs every now and then. As a small boy, I never could get used to those brown eggs. I knew eggs ought to be white on the outside and couldn't believe a brown egg was a good egg.

My advice to you is to stick with the most popular breeds of chickens for your family flock. These have come out on top because they have performed best over a great many years.

Farmers who have breeding flocks and hatcheries concentrate on the most popular breeds. You can get the best quality chicks when you select the top breeds hatched by the millions each year. And you also will get the best prices.

The first consideration is to select the breed that fits your purpose. Here are some suggestions:

Eggs: The Single Comb White Leghorn has been proven the best egg producer. Today, White Leghorns, strain crosses of White Leghorns, and hybrid crosses using a considerable amount of Leghorn parentage make up most of the commercial egg-producing chickens in the U.S. Leghorn males do not make good meat birds and are unprofitable even if the chicks are given to you.

Eggs and Meat: White or Barred Plymouth Rocks, Rhode Island Reds, and New Hampshires are good choices for dual-purpose flocks.

Meat: Crosses utilizing Cornish breeding offer the most economical meat production.

Here are some good rules for chick selection:

- Buy direct from a reputable hatchery that has a U.S. Pullorum-Typhoid Clean rating.
- If possible, work closely with the person from whom you buy the chicks. If you are new in the poultry business, the hatcheryman can help you.
- Neighbors who have had good results may give you the name and address of the hatchery where they buy chicks.
- Almost 100 percent of the egg-laying strains of birds sold are Leghorn-type birds. Brown egg-laying strains are available. Remember: There is no difference between white and brown shelled eggs—except for the shell color—if the birds are cared for in the same manner.

How to select the best type of chicken for your family flock

If you were to go to a chicken show, you would be dazzled by the wide variety of classes and breeds on display. However, only a handful are now used for commercial egg and meat production. Chickens exist in many colors, sizes, and shapes. There are more than 350 combinations of physical features. In order to be able to identify and classify each of these we have established a system of designations known as classes, breeds, and varieties.

A class is a group of breeds originating in the same geographical area. The names themselves indicate the region where the breeds originated, such as Asiatic, Mediterranean, or American. The breeds of chickens in this book are arranged first according to their class, and then alphabetically by breed name within each class. Lesser-known classes, breeds, and varieties are at the end of each section.

Breed means a group, each of which possesses a given set of physical features, such as body shape or type, skin color, carriage or station, number of toes, and feathered or non-feathered shanks. If such an individual is mated to one of its own kind these features will be passed on to the offspring.

Variety means a sub-division of a breed. Differentiating characteristics include plumage color, comb type, or presence of a beard and muffs. Examples exist in almost all breeds. In Plymouth Rocks, there are several colors: barred, white, buff, partridge, etc. In each case the body shape and physical features should be identical. The color is the only difference and each of these colors is a separate variety. Another example is the Leghorn breed where most varieties exist in Single Comb and Rose Comb with all other features identical.

Strains are families or breeding populations possessing common traits. They may be subdivisions of a breed or variety or may even be systematic crosses. However, a strain shows a relationship more exacting than that for others of similar appearance. Strains are the products of one person or one organization's breeding program. Many commercial

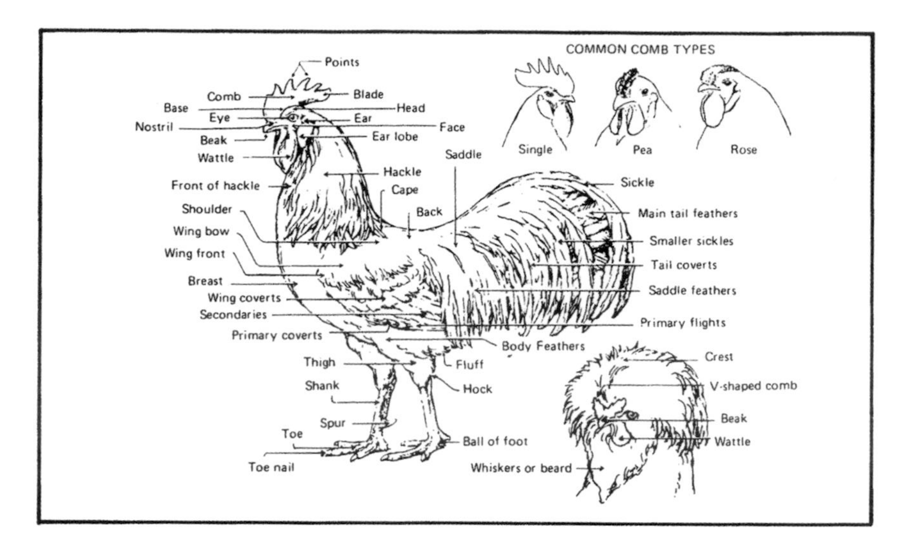

strains exist. Such names as DeKalb, Hyline, Babcock, and Shaver are-organizations that have bred specific strains of chickens for specific purposes.

Most of the breeds and varieties we know in the U.S. today were developed between 1875 and 1925. During that time the emphasis throughout the poultry world was on breeds and varieties. Success was measured in terms of the excellence of individual birds. As the commercial egg and poultry meat industries developed, the emphasis changed from the individual bird to the average for the entire flock. This caused some breeders to adopt intensive selection programs based on the performance of certain outstanding families while others worked with breed crosses and crosses of strains within a breed.

Today the commercial poultry industry is based almost 100 percent on the strain approach. However, foundation breeders are constantly looking for additional material for gene pools. This must come from fanciers and hobbyists who maintain the various breeds for personal and esthetic reasons rather than strictly for the production of meat and eggs.

The American Poultry Association issued a book called *The American Standard of Perfection.* This book contains a complete description of each of the more than 300 recognized breeds and varieties. Such things as size, shape, color, and physical features are described and illustrated in detail.

Leghorns

Varieties: Single Comb Dark Brown, Single Comb Light Brown, Rose Comb Dark Brown, Rose Comb Light Brown, Single Comb White, Rose Comb White, Single Comb Buff, Single Comb Black, Single Comb Silver, Single Comb Red, Single Comb Black Tailed Red, Single Comb Columbian.

Standard Weights: Cock—6 pounds; hen—4 ½ pounds; cockerel—5 pounds; pullet—4 pounds.

Skin Color: Yellow.

Eggshell Color: White.

Use: An egg-type chicken, Leghorns figured in the development of most of our modern egg-type strains.

Origin: Leghorns take their name from the city of Leghorn, Italy, where they are considered to have originated.

Characteristics: A small, sprightly, noisy bird with great style, Leghorns like to move about. They are good foragers and can often glean much of their diet from ranging over fields and barnyards. Leghorns are capable of considerable flight and often roost in trees if given the opportunity. Leghorns and their descendants are the most numerous breed we have in America today. The Leghorn has relatively large head furnishings (comb and wattles) and is noted for egg production. Leghorns rarely go broody.

Plymouth Rocks

Varieties: Barred, White Buff, Partridge, Silver Penciled, Blue, Columbian.

Standard Weights: Cock—9 ½ pounds; hen—7 ½ pounds; cockerel—8 pounds; pullet—6 pounds.

Skin Color: Yellow.

Eggshell Color: Brown.

Use: Meat and eggs.

Origin: Developed in America in the mid to latter part of the nineteenth century. The barred variety was developed first. It was noted for its meaty back and birds with barred feathers brought a premium on many markets. Most of the other varieties were developed from crosses containing some of the same ancestral background as the barred variety. Early in its development, the name Plymouth Rock implied a barred bird, but as more varieties were developed, it became the designation for the breed.

Characteristics: Plymouth Rocks are a good general farm chicken. They are docile; normally will show broodiness; possess a long, broad back, a moderately deep, full breast, and a single comb of moderate size. Some strains are good layers while others are bred principally for meat. White Plymouth Rock females are used as the female side of most of the commercial broilers produced today. They usually make good mothers. Their feathers are fairly loosely held but not so long as to easily tangle. Generally, Plymouth Rocks are not extremely aggressive, and tame quite easily.

New Hampshire Reds

Varieties: None.

Standard Weights: Cock—8 ½ pounds; hen—6 ½ pounds; cockerel—7 ½ pounds; pullet—5 ½ pounds.

Skin Color: Yellow.

Eggshell Color: Brown.

Use: A dual-purpose chicken, selected more for meat production than egg production. Medium heavy in weight, it dresses a nice, plump carcass as either a broiler or a roaster.

Origin: New Hampshires are a relatively new breed, having been admitted to the Standard in 1935. They represent a specialized selection out of the Rhode Island Red breed. By intensive selection for rapid growth, fast feathering, and early maturity and vigor, a different breed gradually emerged. This took place in the New England states—chiefly in Massachusetts and New Hampshire, from which it takes its name.

Characteristics: They possess a deep, broad body, grow feathers very rapidly, are prone to go broody, and make good mothers. Most pin feathers are a reddish buff in color and, therefore, do not detract from the carcass's appearance very much. The color is a medium to light red and often fades in the sunshine. The comb is single and medium to large in size; in the females it often lops over a bit. These good, medium-sized meat chickens have fair egg-laying ability. Some strains lay eggs of a dark brown shell color. New Hampshires are competitive and aggressive. They were initially used in the Chicken of Tomorrow contests, which led the way for the modern broiler industry.

Rhode Island Reds

Varieties: Single Comb, Rose Comb

Standard Weights: Cock—8 ½ pounds; hen—6 ½ pounds; cockerel—7 ½ pounds; pullet—5 ½ pounds.

Skin Color: Yellow.

Eggshell Color: Brown.

Use: A dual-purpose medium heavy fowl; used more for egg production than meat production because of its dark-colored pin feathers and its good rate of lay.

Origin: Developed in the New England states of Massachusetts and Rhode Island, early flocks often had both single- and rose-combed individuals because of the influence of Malay blood. It was from the Malay that the Rhode Island Red got its deep color, strong constitution, and relatively hard feathers.

Characteristics: Rhode Island Reds are a good choice for the small flock owner. Relatively hardy, they are probably the best egg layers of the dual-purpose breeds. Reds handle marginal diets and poor housing conditions better than other breeds and still continue to produce eggs. Some "Red" males may be quite aggressive. They have rectangular, relatively long bodies, typically dark red in color. Most Reds show broodiness, but this characteristic has been partially eliminated in some of the best egg-production strains. The Rose Comb variety tends to be smaller but should be the same size as the Single Comb variety. The red color fades after long exposure to the sun.

Wyandottes

Varieties: White, Buff, Columbian, Golden Laced, Blue, Silver Laced, Silver Penciled, Partridge, Black.

Standard Weights: Cock—8 ½ pounds; hen—6 ½ pounds; cockerel—7 ½ pounds; pullet—5 ½ pounds.

Skin Color: Yellow.

Eggshell Color: Brown.

Use: Meat or eggs.

Origin: America. The Silver Laced variety was developed in New York State and the others in the north and northeastern states in the latter part of the nineteenth century and early twentieth century.

Characteristics: Wyandottes are a good, medium-weight fowl for small family flocks kept under rugged conditions. Their rose combs do not freeze as easily as single combs and the hens make good mothers. Their attractive "curvy" shape, generally good disposition, and many attractive color patterns (varieties) make them a good choice for fanciers as well as farmers. Common faults include narrow backs, undersized individuals, and relatively poor hatches. Also, it is not uncommon to see single-combed offspring come from rose-combed parents. These single-combed descendants of Wyandottes should not be kept as breeders.

Cornish

Class: English.

Varieties: Dark, White, White Laced Red, Buff.

Standard Weights: Cock—10 ½ pounds; hen—8 pounds; cockerel—8 ½ pounds; pullet—6 ½ pounds.

Skin Color: Yellow.

Eggshell Color: Brown.

Use: Developed as the ultimate meat bird, the Cornish has contributed its genes to build the vast broiler industry of the world. Its muscle development and arrangements give excellent carcass shape.

Origin: Cornish were developed in the shire (county) of Cornwall, England, where they were known as "Indian Games." They show the obvious influence of Malay and other Asiatic blood. They were prized for their large proportion of white meat and its fine texture.

Characteristics: The Cornish has a broad, well-muscled body. Its legs are of large diameter and widely spaced. The deep set eyes, projecting brows and strong, slightly curved beak give the Cornish a rather cruel expression. Cornish males are often pugnacious and the chicks tend to be more cannibalistic than some breeds. Good Cornish are unique and impressive birds to view. The feathers are short and held closely to the body, and may show exposed areas of skin. Cornish need adequate protection during very cold weather as their feathers offer less insulation than can be found on most other chickens.

Chapter 3

How to start your home poultry flock

WE HATCHED ABOUT 500 CHICKS every spring on the Iowa farm where I grew up. Two incubators were set up in the basement of our house. Each had its own kerosene heating unit to keep the eggs at the right temperature. Water was placed in trays to maintain the proper humidity.

In late afternoon, about the time I came home from school, Mother turned the eggs. She pulled out each tray of eggs and gave the eggs a quarter turn or so. She kept a sharp eye on the temperature all during the 21-day incubation period.

The day finally came when we would hear the first peep. Then there was the excitement of looking inside to watch a chick hatch. First came a small crack in the shell. Then the bill appeared. The chick hammered away at its prison until it finally broke the shell open. The chick appeared damp and exhausted at first but within minutes it was dry and chirping around. Within 24 hours almost all of the eggs would be hatched ready for the move to brooding.

Looking back, it seems to me that a high percentage of the eggs hatched. It's a tribute to the development of the incubator in those distant days. And I think the purchase of eggs from a reliable hatchery plus good care made a lot of difference. If you decide to hatch your own chicks those same requirements are as important today as you plan your family flock.

There are four ways to get your home poultry flock started. You should choose the one that best fits your experience and your family's desire to be involved in the enterprise. The alternatives are to hatch fertile eggs, buy day-old chicks, buy started pullets, or buy second-year hens from a commercial producer.

Obviously, buying the pullets or the hens is the easiest way to get started. There are fewer risks and much less care is required. You will be eating your own eggs immediately. If you are honest about the costs, this may even be the most economical way to begin.

Most home poultry flock owners want the feeling of accomplishment. They like the idea of seeing the chicks hatch or caring for them at the fluffy yellow down stage. Children enjoy watching the chicks grow and it helps them learn about life. You do need more equipment, more facilities, and lots more time when you start with eggs or the chicks.

If you are interested in broilers or fryers, starting with the eggs or the chicks are your only alternatives. Since the birds are ready to eat in six to eight weeks, buying older birds isn't a viable alternative. A started chick is ready to eat in a month.

Hatching your chicks

The first consideration is the availability of fertile eggs. These must come from a poultryman who is in the business of producing eggs for a hatchery. Eggs you buy from a store are not fertile. Commercial flocks have no males. This is a surprise to the novice poultryman but the fact is that a cockerel is not necessary in the production of eggs for the table.

You can search online for hatcheries. Talk to the county agricultural extension director or ask other poultrymen for possible sources.

If possible, pick up the eggs yourself rather than having them shipped or mailed. It is difficult for hatcheries, the post office, or UPS to handle small orders of eggs properly.

Care of eggs prior to incubation

Fertile eggs are perishable and need careful handling before incubation begins. Avoid extreme variations in temperature. Ideally eggs should not be more than seven days old when they are set. Beyond that point, the percentage that will hatch declines.

If it is necessary to hold the eggs before setting, turn them daily keeping them in a room where the temperature is around 50° and the relative humidity is 70 to 80 percent. The vegetable section of your refrigerator could be used for holding eggs until you are ready to incubate. Temperatures below 40° reduce hatchability. Do not hold at room temperature because embryos begin to develop at subnormal rates when temperature reaches 80°. This will reduce the number that hatch.

How to incubate your eggs

Eggs have been incubated by artificial means for thousands of years. Both the Chinese and the Egyptians are credited with originating artificial incubation procedures. The Chinese developed a method in which they burned charcoal to supply the heat. They also used the hot-bed method in which decomposing manure furnished the heat. The Egyptians constructed large brick incubators which they heated with fires right in the rooms where the eggs were incubated.

Over the years incubators have been refined and developed to the point where they are almost completely automatic. Modern commercial incubators are heated by electricity, have automatic egg-turning devices, and are equipped with automatic controls to maintain the proper levels of heat, humidity, and air exchange. Present-day commercial incubators vary in capacity from a few thousand to many thousands of eggs, and they have made possible the development of modern hatcheries that produce almost all the chickens grown in this country.

This picture shows a typical commercial incubator for the family flock. The unit has an electrical heating element with a thermostat for automatic temperature control. A moisture pan helps maintain humidity.

You can build a small incubator or you can buy a unit from various websites or mail order catalogs. These have been designed to serve the needs of the family flock owner. Built-in heat control and ventilation take some of the risk out of hatching eggs. However, the investment may be more than you can justify for a few birds.

If you have average handyman skills, you can build an incubator suitable for two or three dozen eggs. Plans for two units follow. The cardboard box model was developed by E. A. Schano from Cornell University, and the plywood box was developed by John Bezpa from Rutgers University.

A cardboard box incubator

The following supplies are needed to construct an inexpensive cardboard box incubator that will hold three dozen or more eggs:

- Two cardboard boxes, one 16" wide x 20" long x 12 ½" high, the other 14" wide x 18" long x 13" (or more) high
- Single-strength pane of glass 16" x 20"

- ¼"-mesh welded hardware cloth 18" x 22"
- Commercial heating unit or porcelain socket and light bulb
- Cake tin (water pan) 1 ½" deep x approximately 9" x 14"
- Glue
- Masking or scotch tape
- Newspapers
- Two brooding or incubator thermometers

The incubator is made in the following manner:

1. Place the smaller box inside the larger one. The inner box should be higher than the outer box and approximately two inches smaller in both length and width.
2. Mark a line on the inside box approximately ¼ inch below the level of the outside box. Use a yardstick to make a straight line on the inner box after it is removed from the outer box.
3. Cut off the top of the inside box along the line made in step number 2.
4. Use cut-away pieces of the inside box to line the bottoms of both the inner and outer boxes where the flaps do not meet. If there are no cut-away pieces, use pieces from a third box.
5. Put glue on the bottom of the inner box and then center the inner box in the outer one. There should be a one-inch space between the sides of the boxes. Secure the inner box until the glue dries.
6. Mark a line on the flaps of the outside box where they come in contact with the inner edge of the inside box.
7. Cut off the flaps of the outside box along the lines drawn in step number 6. Cut the corner pieces on a diagonal so that they will make a neat, flat corner.
8. Stuff strips of newspaper lightly into the space between the boxes. Do not bulge the sides of the incubator. Wood shavings, excelsior, or styrofoam can be used in place of the newspaper strips.
9. Use tin snips to cut a two-inch square from each corner of the ¼-inch mesh hardware cloth, then bend the projecting pieces of the screen down so that they form legs to support the screen.

10. Place the cake tin, which will cover about one-half of the surface area of the inside box, under the hardware cloth screen.
11. Install the commercial heating element as directed in the instructions sent with the unit. If you use an electric light for heat, mount the porcelain socket on a board six inches square, then place the mounting board on the screen. Next, place a tube of cardboard around the light. Position the tube so that it surrounds the light and stands like a chimney, but to reduce the fire hazard do not let it come in contact with either the lightbulb or the covering glass. An oatmeal box makes a good tube.
12. Tape the flaps of the outer box to the sides of the inner box. This seals the area in which the insulating material was placed.

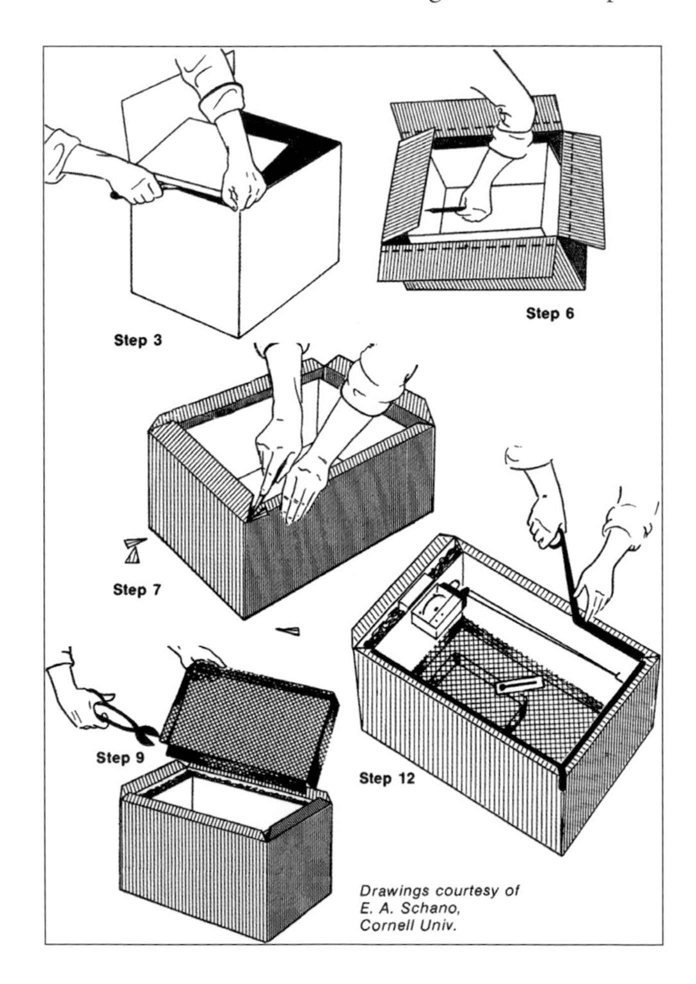

Drawings courtesy of E. A. Schano, Cornell Univ.

A plywood incubator

The following materials are needed:

- Plywood
- Glass
- Cake tin
- ¼"-mesh welded hardware cloth
- Heating unit: either a commercial unit or porcelain socket and lightbulb
- Masking tape

You may construct the incubator according to the size desired. It can be a small one for only a few eggs, or it can be a somewhat larger one that will hold several dozen eggs. The larger the incubator, the more difficult it will be to maintain a uniform temperature in it. In fact, you may find it important in the larger incubator to put in a small fan (three-or four-inch blade or smaller) with a low revolving rate.

Don't expect 100 percent success in hatching eggs in these or any other incubators. Commercial hatcheries with all their highly automatic and specialized equipment do not average much more than an 80 percent hatch of all the eggs they incubate. You probably should not count on hatching more than 50 percent, and you may not even succeed in hatching 50 percent.

Location of incubator

To help your incubator maintain a constant temperature, place it where it will receive as little temperature fluctuation as possible. Do not place it near a window where it will be exposed to direct sunlight. The sun's heat can raise the temperature high enough to kill the developing embryos. Connect the unit to a dependable electrical source, and make sure the plug cannot be accidentally detached from the outlet.

Preparing the incubator

Before you incubate, be sure the incubator is working properly and that you know how to operate it. Place warm water into the humidity pan, and adjust the heat source until the incubator temperature stays between 99° and 102° F. Check the thermometer frequently for at least 24 hours before you incubate to be sure it will stay at the correct temperature.

Heat sources having a thermostat are most reliable. Lightbulb units without a thermostat can be difficult to control unless the room temperature is relatively constant.

Occasionally, people attempt to incubate eggs in ovens or other unconventional facilities. They are nearly always disappointed because temperature and humidity requirements are within a quite narrow range.

Even with good equipment, incubation is not always successful, so make every attempt to provide the proper environment—using a reliable incubator.

When eggs are placed into an incubator operating at the proper temperature, the temperature will drop. Do not adjust the thermostat upward during this warmup period. The time that the temperature in the unit will remain below normal depends upon the temperature of the eggs and the capacity of the heating unit. This temperature lag period can be reduced by warming the eggs to room temperature before they are placed into the incubator.

Incubator operation

Temperature—Maintain the temperature in the 99° to 102° F. temperature range (100° to 101° F., if possible). Place the thermometer to measure the temperature at a level at or slightly above where the center of the egg will be. Overheating the embryo is much more damaging than is underheating it; overheating speeds up embryo development, lowers the percentage of hatchability, and causes abnormal embryos. Although a short cooling period may not be harmful, longer periods of low temperatures will reduce the rate of embryo development. Excessively low temperatures will kill the embryos. Avoid temperatures outside the 97° to 103° F. range. If the temperature remains beyond either extreme for several days, hatchability may be severely reduced.

Humidity—The moisture level in the incubator should be about 50 to 55 percent relative humidity, with an increase to about 65 percent for the last three days of incubation. Moisture is provided by a pan of water under the egg tray. The water surface should be at least half as large as the surface of the egg tray. Add warm water to the pan as necessary. If more humidity is needed, increase the size of the pan or add a wet sponge. Humidity adjustment can also be made by increasing or decreasing ventilation.

Using a wet bulb thermometer, you can determine relative humidity from this chart:

Temperature in degrees F	Wet bulb reading in still-air incubator					
100	81.3	83.3	85.3	87.3	89.0	90.7
101	82.2	84.2	86.2	88.2	90.0	91.7
102	83.0	85.0	87.0	89.0	91.0	92.7
Relative humidity %	45	50	55	60	65	70

If a wet bulb thermometer or hygrometer is not available, the size of the air cell in each egg can be used to estimate whether the humidity should be increased or decreased (see candling instructions). The air cell increases in size during incubation at a rate that depends on temperature and humidity as moisture evaporates from the egg. The drawing shows the normal size of an egg's air cell at 7, 14, and 18 days of incubation.

Ventilation—Ventilation is adjusted by increasing or decreasing openings in the sides or top of the incubator. Normal air exchange is needed during embryo development and should be increased as the chicks begin to hatch. The embryo needs oxygen and produces carbon dioxide. However, the correct relative humidity must be maintained until

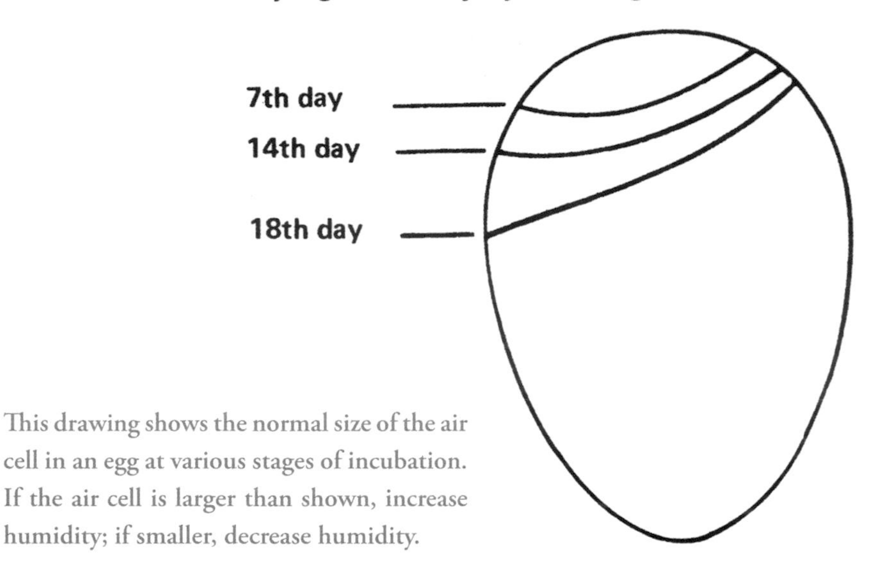

This drawing shows the normal size of the air cell in an egg at various stages of incubation. If the air cell is larger than shown, increase humidity; if smaller, decrease humidity.

most of the chicks are out of their shells. Do not open the incubator unless necessary during the last three days of incubation.

Turning—The eggs should be placed into the incubator on their sides. Turn them at least three times a day, except for the last three days when they don't need turning. Turn the eggs an odd number of times so the position that is up the longest (at night) will be changed from day to day. Mark the date or an "X" on each egg so you can tell if the eggs have been turned. When you turn the eggs, move them to a different part of the tray to minimize the effects of temperature variation in the incubator. If the eggs are not placed on their sides, they should be placed at an angle so the small ends are in the downward position. Weekends often pose a problem for teachers incubating eggs at school. They sometimes take the incubators home (placing the eggs into egg cartons and wrapping them to keep them warm when traveling back and forth), especially the first weekend. Some teachers have found leaving incubators unattended during two-day weekends has had little affect. Often the effect of not turning the eggs may be much less than that of the jostling, jarring, and possible temperature changes involved in taking the eggs home.

Length of incubation

Approximate incubation periods for commonly hatched poultry and game bird species in small incubators are:

Species	Days
Chicken	21
Most ducks	28
Muscovy ducks	33–35
Turkey	28
Most geese	29–31
Ringneck pheasant	23–24
Japanese quail	17–18
Bobwhite quail	23
Chukar partridge	22–23
Guinea fowl	26–28
Peafowl	28

Candling the eggs

"Candling" is the examination of the contents of the eggs using a shielded light in a darkened area. Eggs should be checked for development; then, if fertility is poor, you do not have to wait the entire incubation period to learn you are going to have a poor hatch. Candling to check air cell size can determine incubator humidity. You can also observe the development of the embryo.

You can make an egg candler from a wood or metal box or from a container in which you mount a 40-watt light bulb. Make a one-inch

hole in the end near the bulb. For better viewing, place a felt or cloth cushion around the opening so an egg fits the opening better and so light does not leak around the egg.

Hold the large end of the egg up to the candling light. You won't see much development until the fourth or fifth day of incubation. White or light-colored eggshells permit better viewing of embryo development. The contents of the egg have a pinkish color or cast when the embryo is developing properly. As the embryo grows, it occupies most of the space within the shell. Toward the end of incubation, the contents will appear dark except for the air cell. Eggs that appear clear at 4 to 5 days in incubation or that show little development at 10 days should be removed from the incubator. They are infertile or contain early dead embryos.

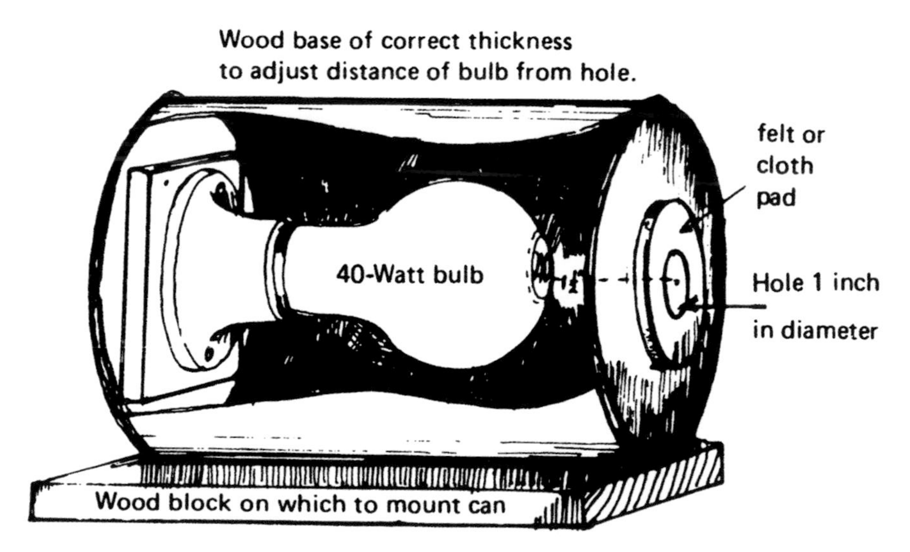

This egg candle can be made from a tin can that's about five inches in diameter and seven to nine inches long. (A shortening can with an easily removable lid works well.)

Candling will not influence embryo development if you handle the eggs gently. When eggs are removed from the incubator only a few times and are not allowed to cool to any extent, candling makes little difference in hatchability or the time required for hatching.

Getting ready for the hatch

When the eggs are last turned, three days before hatch, place a layer of crinoline or cheesecloth on the screen under the eggs. It will make cleaning the incubator easier after hatching.

Most chicks should hatch within a 24-hour period. Late-hatching chicks may lack vigor or be abnormal. After the chicks have dried and fluffed up completely, they can be removed from the incubator. When most of the chicks have hatched, you can lower the incubator temperature to about 95° F. if the chicks are to be kept in the incubator for one or two days.

Clean the incubator after the hatch, so it will be ready for the next time. Sanitation is an important part of incubating. Remove and dispose of the crinoline or cheesecloth, together with the shells and other remains. Clean the inside of the incubator with soap and water, and let it dry completely before putting it away.

Starting with chicks

If possible, order your chicks from a local hatchery several weeks in advance. A personal visit to the hatchery gives you the opportunity to learn about the background of the chicks.

Ask about egg production of the breeding stock. You can't get top production from a poorly bred hen. She must have a high production potential in her make-up. A good goal is 65 percent, or 237 eggs a year per hen housed.

Choose a strain with good egg size and quality. Both are basically inherited. However, there are limits as to how far one should go with egg size. Eggs can be so large as to cost more to produce and so small as to bring lower prices. After mature egg size is reached, the ideal weight is 25 to 26 ounces a dozen. Choose a strain that will lay over 90 percent Grade A or AA eggs. You want birds that are known to have good livability under reasonable environmental and management conditions. The high cost of growing a pullet to maturity makes laying house mortality especially serious. Adult mortality should not exceed 1 percent a month, and growing-period mortality should not exceed 5 percent.

Plan to pick up the chicks yourself. It is amazing how much abuse day-old chicks can stand in shipping. When a chick hatches, nature has provided it with nutrients for about three days. However, elimination of chilling, overheating, and starvation is sure to help your birds get off to a better start. Be sure that mail order chicks are guaranteed.

Baby chicks usually leave the hatchery when they are one day old. They normally require heat during the early weeks. They sometimes are vaccinated, dubbed, sexed, and debeaked at the hatchery.

Straight-run chicks cost less than other ages and types of live birds. Straight-run birds are boxed at random as they come from the incubator; normally, about half the chicks are pullets and half are cockerels. When space and labor are available on the farm, cockerels of medium-weight breeds may be grown as meat-type birds at the same time that pullets are grown as layers. Leghorn cockerels aren't worth the feed they eat and should not be purchased.

Sexed chicks are sorted after they leave the incubator into lots of pullets and cockerels. Sexed pullets of lightweight breeds cost more than twice as much as straight-run chicks; sexed pullets of medium-weight breeds—while usually not as expensive as lightweight breeds—cost more than straight-run chicks.

The average family flock should be started with 25 to 35 straight-run chicks or 15 to 20 sexed pullet chicks. This should give you 12 to 15 pullets at 22 weeks of age. The meat type straight-run birds will give you 12 to 15 cockerels for your table or for sale as well as the layers. Late spring or early summer probably is the best time to start a home poultry flock.

Buy started pullets

Many poultrymen make a business of raising pullets for sale to laying flocks. They either sell immediately after brooding or when the birds are 16 to 20 weeks of age and ready to lay.

Buying pullets eliminates both hatching and brooding. You save on facilities and the amount of time and labor required is vastly reduced.

There should be an understanding between the buyer and the grower concerning the strain or cross of the started pullets, the type and number of vaccinations, the kind of feed, and the disease history of the birds. There are three basic things that will help get satisfactory results when you buy started pullets: (1) Always buy from a person who has a reputation for furnishing a good quality pullet. (2) Place your order for pullets five months or more in advance. (3) Have a written agreement so that both parties fully understand all the little details that should be a part of any buy-sell agreement.

Buy pullets from a hatcheryman or grower who makes a specialty of raising pullets to sell—a person who raises only birds of one age in the same building—and, one who cleans up between broods and gives the buildings a chance to air out. If the pullet raiser is close to home, so much the better. The buyer can drop in occasionally and see how the pullets are coming along. Also, a short haul for started pullets is better than a long one.

Ask the same questions about started pullets as you would in buying chicks. What is the background of the parent stock? How well is it adapted to your climate and equipment? How about resistance to common diseases?

One of the advantages of started pullets is that they usually have been vaccinated and debeaked to avoid feather picking and cannibalism.

Started pullets usually are sold between six and eight weeks of age. Individual lots may vary with seasonal needs for heat. Their high purchase price reflects the costs involved in getting the birds through brooding.

Ready-to-lay pullets are sold at 16 to 20 weeks of age. They begin to lay almost as soon as they reach the farm. Ready-to-lay pullets sell for more than any other age group of similar quality birds because they are through their unproductive months. The grower has invested considerable time and feed in them.

Start with second-year layers

Second-year or "recycled" hens offer the easiest, most economical way to establish the home poultry flock.

Large commercial flocks usually replace their layers after 12 to 15 months of production. Pullets begin laying when they are 20 to 25 weeks of age. Production peaks at 34 weeks, then declines steadily at about 0.5 percent per week until the hens molt. During the molting period, production drops drastically. After this relatively short phase is over, the birds return to a laying rate perhaps 15 percent below the first year.

The lower rate may not be profitable for the large, automated flock. However, it works out well for the home poultry flock. The eggs are large and have good shell quality.

The purchase price will vary with age, breed, and strain of the birds and their potential productive life. You'll find them very cheap compared to started pullets. However, make sure that the hens are in the 12- to 15-month age range. A bird heading into her third year is a poor choice. You should neither buy a hen of that age nor keep her in your own flock.

When a hen molts, she loses her old feathers and grows new ones. Egg production normally stops as molting starts. The hen first sheds head and neck feathers. She then goes on to lose feathers in this order—breast, body, wing, and tail.

The birds in a flock that molt first are the low producers. Try to avoid them in selecting hens for your home poultry flock. The early molters may not resume satisfactory egg production for four to six months. In contrast, the birds that molt late usually are high producers. After a rest of two or three months, they begin to lay again.

Chapter 4
Brooding the chicks

ONE OF THE HAPPY SIGHTS of farming is a crowd of young chicks happily scratching around the edge of the hover. When temperatures are just right, the fluffy chicks peep continually as they scamper from food to water and back again. It's a cheerful sound. When it is chilly or drafty, the tone changes and the chicks let you know something is wrong.

The heat lamps and other brooder units now available take most of the worry out of brooding chicks, particularly in small batches. Back in my youth on the farm, brooding chicks was a tricky business.

Our first brooder stove was fired by coal. It was a bear to keep at an even temperature. Someone had to stoke the fire the last thing at night and it was the first chore in the morning. There may have been midnight patrols on cold nights. After some problem filled the house with smoke and killed a lot of chicks, we shifted over to an oil burner. What a joy it was to have dependable heat with an automatic control.

It's amazing how fast chicks grow. Scarcely more than a week after brooding begins, the first wing feathers begin to appear. Within a month feathering is well along. The cute chick has become an awkward adolescent. The hover can be raised and the heat turned down. Those little chicks have a mighty strong will to live and their progress toward maturity is amazing.

Newly hatched chicks are at a critical stage of life. They require warmth, protection, and the right feed during those first few days. While temperature requirements are quite strict, it doesn't take a fancy building to handle a family-size batch of chicks.

A small outside building with a floor and solid construction will do. Many chicks are started in a partition across the back of the garage. It is not wise to start your chicks in the basement or a room inside the house. Chick odor and down may hang around the house long after the birds are moved out.

A large cardboard or plywood box is all you need to start. It is better than keeping chicks on the floor of a brooder house where disease from previous flocks may linger. A box 2 or 3 feet wide by 3 or 4 feet long and 15 to 20 inches deep will house 25 to 35 chicks for 2 or 3 weeks.

A hardware-cloth cover on the box will confine the birds as they grow and learn to fly out. Use clean dry litter in the box. Wood shavings are good. Rice hulls, shredded cane, sphagnum peat moss, or dry sand are other possibilities. Availability will be a consideration. Avoid coarse or moldy materials; and especially at first, avoid any slick surfaces. Chicks can be started on wire. Hardware cloth (1/2" x 1/2") nailed to a 1" x 2" or 1" x 3" wooden frame is best for young chicks. Put litter or paper under the wire for easy regular cleaning.

If chicks are started on the floor of a larger room in warm weather, instead of in a box as already described, be sure to use a chick guard to keep them near the heat, feed, and water for a week or so. The guard can be a 12- to 15-inch strip of corrugated cardboard or wire netting, formed in a circle, which may be enlarged to give more room. By the time chicks can fly over the guard it can be removed.

Brooding, feeding, and watering equipment can be purchased from local feed and farm supply outfits and websites, or through mail order catalogs. Much of your equipment can be built at home.

Brooders available for sale include electric, gas, and oil-fired models. These usually include thermostats to control the temperature.

A simple brooder can be made using an infra-red bulb. Make sure it has fire safety features such as porcelain sockets, chains to adjust the height, and tip-up guard hoops.

A 250-watt bulb is plenty of heat for the 25- to 35-bird flock we have been discussing. If the weather is mild, you may be able to get by with an ordinary 60- to 100-watt bulb placed inside a tin can shield. The 250-watt infra-red lamp will consume six kilowatt hours of electricity per day. The lightbulb takes less. Hang the bulb about eight inches above the floor.

If the average brooder house temperature is 50° F., one 250-watt infra-red lamp is generally sufficient for heating 80 chicks. One chick can be added to this estimate for every degree over 50° F.; one chick should be subtracted for every degree below 50° F. The use of more than one lamp is recommended so chicks will not be without heat if a lamp burns out. Use two 125-watt lamps instead of one 250-watt. Supply more heat by lowering the lamps to 15 inches above the litter, or use more or higher wattage lamps. To reduce heat, turn off some lamps, use smaller lamps, or raise the lamps to 24 inches above the litter. You are heating the chicks only and not the air, so air temperature measurements cannot be used as a guide to chick comfort when using infra-red lamps.

When using a brooder, start the chicks at 90° to 95° F. measured two inches off the floor under the edge of the hover. Reduce the temperature 5° per week until supplemental heat is no longer needed. Watch the chicks as a guide to their comfort. If the chicks crowd together under the brooder, increase the heat. Lower the temperature if they tend to move away from the heat source. Allow 7 to 10 square inches of space under the brooder for each chick. Start the brooder the day before the chicks arrive and adjust to proper operating temperature.

Keep light burning

Lighting programs will vary depending on whether the chicks are to be grown for broilers or layers. In either case, have all-night lights in the brooder house for the first three weeks. These lights do not need to be bright, but they should provide sufficient light for the chicks to find their way around the pen.

Chicks show when they are comfortable

Chicks indicate when they are too cold or too warm. When too cold, they chirp and complain a lot; when too hot, they will lie down or try to pile in corners. When comfortable, young chicks form a ring on the floor under the heat source.

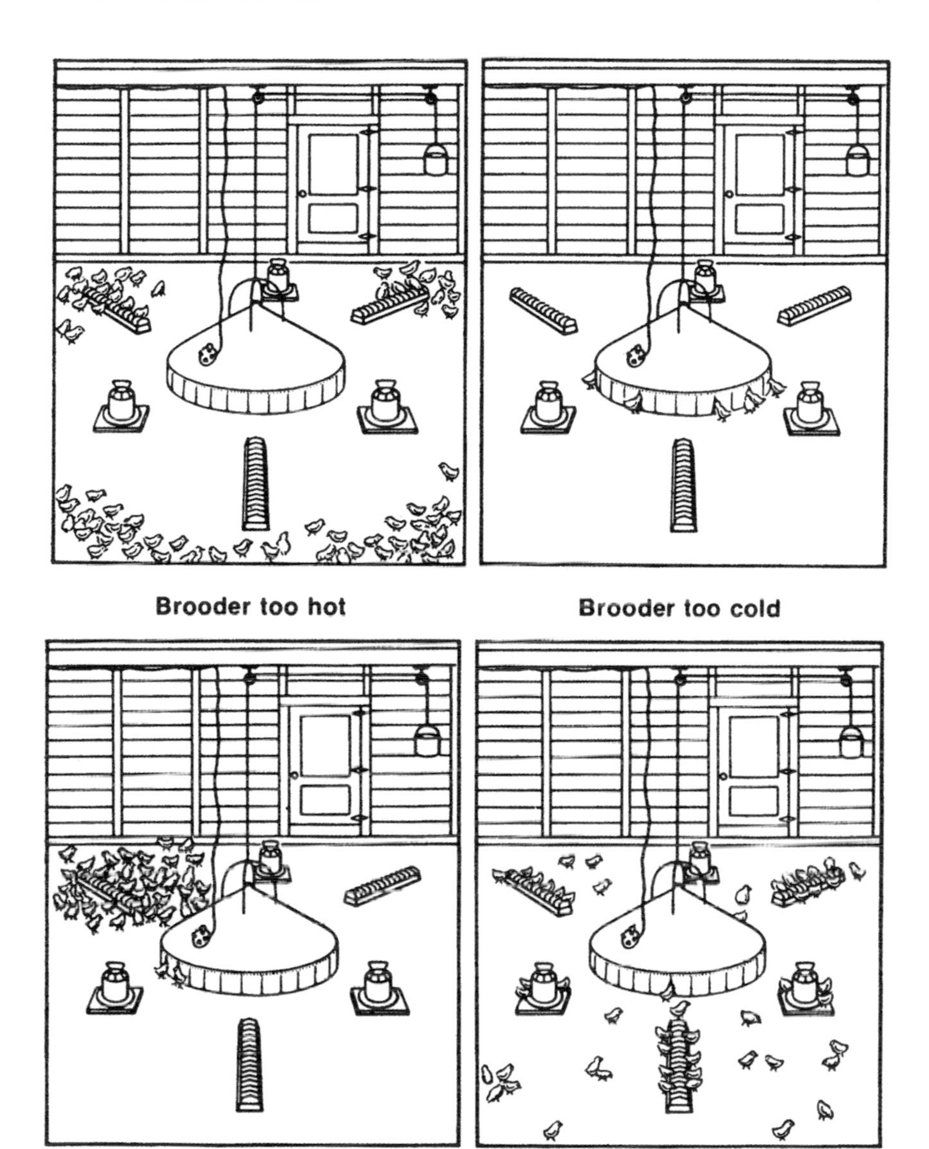

Brooder too hot

Brooder too cold

Brooder house drafty

Chicks are comfortable

Brooding and feeding equipment

Brooding and feeding equipment

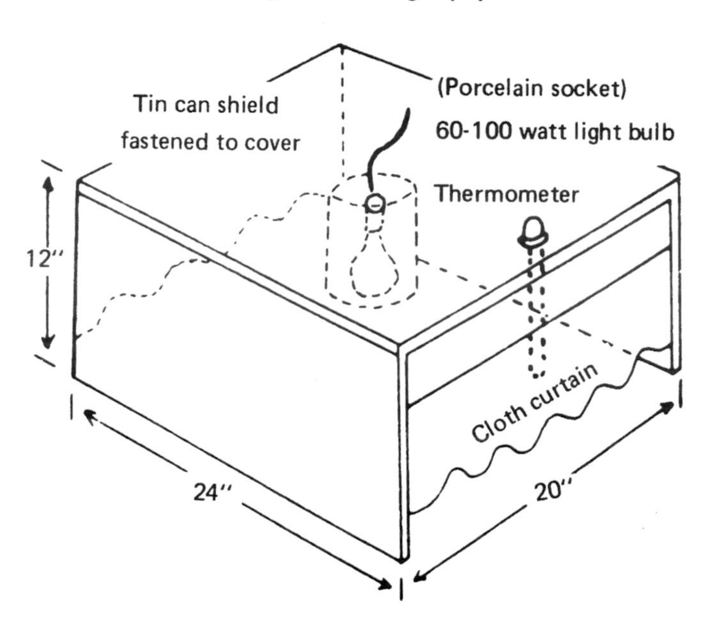

Homemade brooder for 25 to 50 chicks

Cardboard corral with heat lamp

Drawings courtesy of Univ. of Minnesota

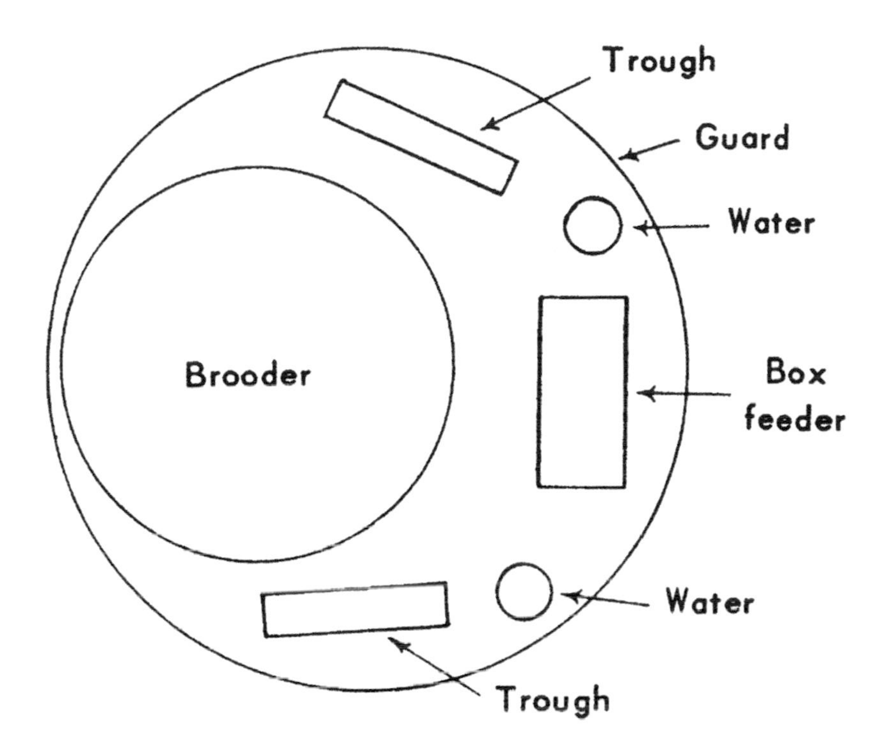

Provide plenty of feed and water

Have both feed and water available from the minute you place chicks in the brooder area. Place feed on chick box lids or trays from cut-down cardboard boxes for the first few days. Provide one lineal inch of feeding space per chick at the hoppers at the start and increase to about two inches after the chicks are two weeks old. After eight weeks provide three to four inches of feeding space for growing pullets. A hanging tube-type feeder 15 inches in diameter can be used for feeding about 30 birds. Less feed is wasted by having hoppers only about half full and adjusting feeder height or size to bird size.

Chicks will waste feed if the trough is overfilled. Keep the feed trough full during the first week only, three-quarters full the second week, and no more than half full thereafter. Adjust the height of the feeder so that the top edge is equal to or slightly higher than the bird's back. This will reduce feed wastage and help prevent litter accumulating in the feed trough. The chicks should be allowed to eat all of the feed out of the trough at least once each week. This will prevent old feed from accumulating in the trough. Remove all wet or moldy feed as this can cause disease problems.

From day-old to four weeks of age, provide two half-gallon water fountains for up to 50 chicks. After the fourth week add another half-gallon fountain. Water fountains should be placed on a wire platform about two inches high. This will help prevent water spillage and keep litter out of the waterer. Clean the water fountain and fill with fresh water daily. If automatic waterers are to be used, start them as soon as the chicks are able to drink from them, but do not remove other fountains until all the chicks have started drinking from the automatic waterers. Adjust the height of automatic waterers in the same manner as described for feeders.

What to feed your chicks

A commercial starter mash purchased from a feed store contains everything your chicks need. Start with a 22 percent protein mixture. The table on the next page shows what you will need.

Actual protein levels of purchased feeds vary somewhat between suppliers. Consumption rates can vary markedly, depending on the feed used, climatic conditions, general management, and age at slaughter or production levels.

The complete ration is designed to fulfill all of the nutritional needs of the bird. However, the home flock owner is usually not highly concerned with getting the best possible performance. In such cases, some savings in feed costs can be obtained by feeding other available materials such as home-grown grains, green plants, and some table scraps.

Chickens	Age	Type feed	Protein	lbs. per 25 birds*
Broiler/Fryers	0–4 weeks	starter	22	50
	5–7 weeks	grower	20	125
	to 8 weeks	finisher	18	25
Roasters	0–4 weeks	starter	22	200
	4–12 weeks	grower	16	1200
	to 15 weeks	finisher	14	1000
Pullets	0–8 weeks	starter	20	100
	8–12 weeks	grower	17	100
	12–21 weeks	developer	14	250

*Pounds consumed in period indicated.

Keep pace with chick growth

You'll need more living space, feeders, and waterers as the chicks grow. After three weeks it will be time to add larger hoppers and waterers. During the first three to four weeks, one-half square foot per bird is needed; during the next four weeks, you'll need one square foot or twice as much.

The chicks rapidly outgrow their box brooder and you need to plan ahead for growth. At three to four weeks, the birds are about half feathered. Heat can be reduced. Your goal is to get the chicks ready to move into the rearing phase in an unheated brooder house.

Chapter 5
Rearing the laying flock pullets

PULLETS ARE THE CUTE ADOLESCENTS of the poultry world. They resemble junior high school students in a lot of ways. They enjoy showing off their new maturity and independence. Pullets are social birds who like to gather together at the water fountain and the cafeteria.

On our farm, we turned the pullets loose into their grassy range during the warm days of May. Feeders and waterers went outside. It only took a week or so for the birds to explore their three-acre "park." As adventuresome as they were during the day, they still liked to return to the safety and warmth of the brooder houses at night. As the summer wore on, more and more of the young birds began to roost in tall evergreen trees near the houses. Many of those Leghorns could fly twenty-five yards or more.

The birds ate like adolescents too. They needed a lot of grain and protein to fill out their rapidly growing frames. And on those hot summer days, bucket after bucket of water was needed to keep the waterers filled.

We raised straight-run chicks and left the cockerels with the pullets. Around the Fourth of July, the males began to show off for the girls. Starting with a few tentative attempts, they soon learned how to give the full notes of the rooster crow. Their favorite time to practice was the hour just after dawn when hard-working farm boys would prefer to sleep.

When chicks are well-feathered they are ready to begin the next phase of growth called rearing. This is the stretch that lasts until the birds are ready to lay at 20 to 22 weeks of age.

There are several alternatives. If your brooder house is large enough, the birds can continue their growth there. You need about two square feet per bird.

If you are starting your first group of pullets, they can go directly into the laying house. However, if you already have a set of hens, you'll want to keep them laying eggs until replacement pullets are ready. Don't mix birds of different ages. The younger birds can't compete for feed and they are exposed to disease.

Another alternative is a cage built off the ground. Birds are grown on a wire mesh floor and are sheltered with a solid roof.

You may want to rear pullets outside on grass or some other forage. Not many suburban home flock owners have the space. If you are going to use a range, it must be fenced six feet high. Your home poultry flock of 25 to 35 birds will need 3,000 square feet of range.

A favorable environment for started pullets may be provided either in a poultry house or on range; healthy, vigorous replacement birds are grown under both systems.

The trend is to raise and to keep laying flocks entirely in confinement. Large pole-type houses are widely used for rearing birds in commercial flocks.

Confinement takes less land and labor than range-rearing systems. Confined birds show fewer losses from coccidiosis, from worms and other parasites, and from such natural enemies as hawks, crows, and foxes. Although cannibalism often is a problem in confined flocks, it usually can be prevented or controlled by debeaking.

Range-grown pullets mature a few days later than confined birds. They sometimes have trouble adjusting to the limited space of the laying house.

Confinement system

Birds may be kept in the same house in which they are brooded, or they may be transferred to a growing house when they are well feathered. Purchased started pullets usually are placed directly in a growing house.

Some extra space may be obtained by careful culling of the flock when birds are transferred to the growing house, or when cockerels are separated from the pullets, at about six weeks. Cull only obviously unhealthy or unthrifty birds from bred-to-lay stock.

Keep litter dry, and add new material as necessary. When roosts are not used, litter requires extra care. Parasites—especially lice—must be controlled.

Perform the stress-producing jobs usually associated with moving pullets—vaccinations, debeaking, dusting for lice and mites—while birds are still in the growing house.

Install a few nests in the growing house before pullets start laying.

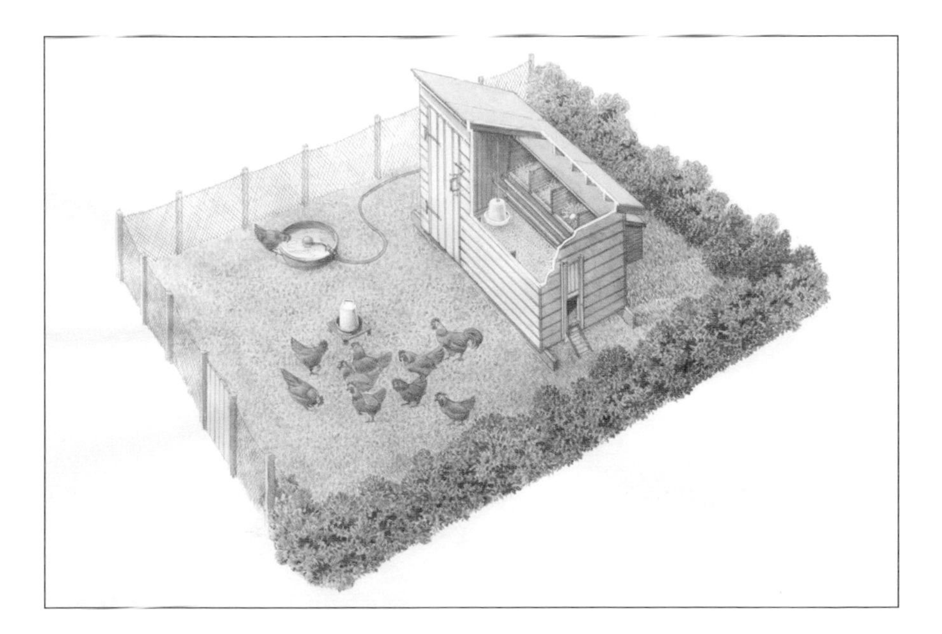

Range system

Range-reared birds benefit from exercise, sunlight, and green feeds. Range may be used for pullets from the time they are brooded until they are ready for the laying house, or for cockerels to be used for breeding.

Birds on range may eat 5 to 15 percent less mash and grain than confined birds. Range rearing requires more labor than the rearing of confined birds; parasites and predators on range are major problems.

Often, it is difficult to find clean land for range. To avoid outbreaks of disease in range-reared chickens, clean range should always be used.

Clean range can be maintained by using a two- to three-year range-rotation program. At the end of a growing season after mature birds have been removed, the range is planted to pasture grass or cover crops. Droppings from the poultry house should not be used to fertilize the area.

Allow the birds to run freely. Do not crowd the range, and do not mix birds of different ages in one area.

Cannibalism

Toe-, feather-, or body-picking may occur in confined flocks, even when birds are properly managed and well fed.

Toe-picking may be the first sign of cannibalism. Remove injured chicks immediately and paint their toes with a stop-pick preparation. Allow injuries to heal before returning chicks to the flock.

Debeaking may be used either to halt picking or to prevent it. Some poultrymen debeak routinely; others practice it only when cannibalism appears.

Chickens can be debeaked at any age. Remove about two-thirds of the upper beak and one-third of the lower beak with a heated blade. The cut should be cauterized to prevent regrowth, bleeding, or infection. A special instrument known as a debeaker is used by commercial poultrymen.

If previous flocks have engaged in cannibalism, it is a good idea to order chicks debeaked at the hatchery. Early debeaking may last only 10 weeks.

When pullets are debeaked at 16 to 20 weeks (and at least two weeks before they are moved into the laying house), cannibalism normally is prevented for the rest of the birds' lives.

Debeaking does not damage the health or reduce the vigor of birds.

Vaccination

Effective vaccines have been developed for four major respiratory diseases: Newcastle disease, bronchitis, fowlpox, and laryngotracheitis.

Plan a vaccination schedule to cover the flock from the time the chicks arrive until they complete a year in the laying house. Base the vaccination schedule on the needs of the individual flock and on local conditions.

It is desirable to have vaccinations completed at sixteen weeks, or about one month before pullets are ready to lay. Under normal conditions, pullets should not be vaccinated as late as 20 weeks; this will delay egg production. Revaccinations are necessary if hens are to be kept for a second year of production.

Normally, a vaccine is most effective when it is administered alone. Stress from multiple vaccinations may cause severe losses. In spite of this, some poultrymen give simultaneous vaccinations for two diseases to save time and labor in handling birds.

Birds should be healthy and vigorous at the time they are vaccinated. Losses may be high if the flock is sick or has severe parasitic infestations. Do not vaccinate birds that have coccidiosis. If this disease appears, wait until the birds have recovered before giving them any type of vaccination.

Follow the manufacturer's instructions carefully. Store vaccine at proper temperatures, use fresh supplies, mix as directed, and administer exact dosages. If vaccines are misused, they may fail to protect the flock.

Provide plenty of good water

Most home poultry flocks are growing to maturity during the hottest part of the year. Keep clean, pure water before the chicks at all times. Your flock of 25 to 35 pullets will need 1 ½ gallons per day in moderate weather and 2 or more when it is hot.

Thoroughly clean water fountains or troughs once a day or more often, if necessary. Refill with a clean, fresh supply of water. Be careful not to spill or empty water on litter.

Feed management

To maintain healthy birds, keep fresh feed available at all times. Limit the amount of feed in feeders to the extent necessary to avoid waste. It is a good practice to fill hanging feeders only three-fourths full, and trough feeders only two-thirds full. For efficient feeding, keep the lip of the feeder pan in a hanging tube-type feeder at the level of the birds' backs.

Fill non-automatic trough feeders in the early morning, and during the day whenever feed supplies get low. If leftover feed is not clean and palatable, remove it before refilling feeders. Never put moldy or contaminated feed in feeders. Clean feeders as needed.

Keep a close check on birds' weight and their feed consumption. A drop in feed intake usually is the first indication of trouble—a disease outbreak, molt, stress, or poor management. If the reason for the drop in feed consumption is not readily apparent, consult a poultry specialist.

Keep feed as fresh as possible. Order for frequent delivery—if possible, every two weeks.

Store feed carefully, in a dry, rat- and mouse-proof place, where it will not be subject to damage from moisture or losses from rodents. A large galvanized garbage can with a tight lid makes an excellent storage container for your feed.

Use a growing ration

Your feed supply store can provide you with a growing ration that contains everything your chicks need to grow into productive hens. It may cost more than mixing yourself, but bagged feed mixed at a mill has many advantages. From 6 to 14 weeks, the ration should contain 17 percent protein. From 15 to 20 weeks, 14 percent protein is sufficient.

You can supplement the mash with grain. This will reduce the overall cost, particularly near the end of the rearing phase.

Pullets may begin to receive grain as soon as they start eating growing mash. Corn, wheat, barley, oats, millet, grain sorghum, or combinations of these may be used.

Begin with 10 pounds of grain for each 100 pounds of mash. Increase grain until pullets are getting equal parts of mash and grain. Put grain and mash in separate hoppers.

When pullets are 18 to 20 weeks old, gradually withdraw the growing mash and replace with laying mash over 2 weeks.

Feeding birds on range

Range cannot provide a complete diet for birds. Pullets that get the green feed of the range need the additional nutrients of a growing ration. Mash or pellets usually are fed in one hopper and grain is fed in another.

Some poultrymen use pellets for range feeding, because the larger particles are less subject to blowing out of feeders.

You need about four inches of feeder space per bird. Plan to have enough so all birds (of any age) can eat at the same time. Feeders and waterers should be raised as the birds get older. The top of the feeder side should be raised to at least the level of the bird's back as it stands (in a normal position) on the floor. The birds should have to reach up and over the edge of the feeder. This will help prevent feed wastage.

Your first egg!

One day around the twentieth week, you'll suddenly find an egg on the floor of the confinement house. Or perhaps it will sparkle in the grass of the range.

Don't be disappointed in that first egg. Pullets normally lay small eggs at first. They are just like any normal egg but smaller.

The first egg is a sign that your flock now is mature. The long weeks of waiting are over. It's time to get them housed.

Chapter 6
Managing for more eggs

GATHERING THE EGGS WAS ONE of the first jobs I was assigned on the farm. They didn't trust me with the big morning pickup. My shift was evening when the number of eggs matched the ability of a small boy to carry them.

There were two frightening things about the job. First of all, when I opened the door the hens flew all over the place. It was like an explosion. They never could get used to a person of my size. Then there were the broody hens. They could see I was easy to buffalo. It only took a peck or two on my timid hand to win the right to sit on the eggs until stronger spirits took over in the morning.

We kept our eggs in the basement and Mother put them into one-dozen trays fitted into the egg case. At least once a week we took them to town where they were sold to the local grocer. When he got a load, he took them to Des Moines to a warehouse. The eggs got their first refrigeration at nearly two weeks of age.

It's no wonder that almost everyone who could produced their own eggs. Today, there's still a long trip from egg factory to supermarket. Eggs are refrigerated all the way but there's still nothing that can match the egg laid today by your own hen.

The first few eggs scattered around the brooder house or on the grass outside are a clear signal that it is time to get your pullets into their laying quarters. This is a critical time when good management is needed to bring the young birds to maturity as layers.

Today's laying hens are marvelous machines for converting feed and water into eggs. A good hen (with proper care) will consume about 90 to 95 pounds of feed in 12 months of egg production. For that feed and some 16 gallons of water she will lay approximately 240 eggs. At an average weight of 24 ounces per dozen this is 30 pounds or over 6 times the body weight of the hen. This is truly a remarkable feat. To do this, however, the hen must have good management, including housing, feeding, sanitation, disease prevention, predator control, parasite control, and disposal of wastes and dead carcasses.

Prepare the laying house

Have the laying house ready for its new tenants when your pullets reach 20 weeks of age. This means disposing of last year's birds if you plan to use the same house. Clean out the building and disinfect the floor, roosts, nests, feeders, and waterers. Your pullets will develop enough problems of their own without inheriting the parasites and diseases of the previous flock.

Laying birds perform best when confined to a comfortable, well-ventilated house. The size depends on the number of birds to be housed and the breed of chickens. Allow 1½ to 2 square feet of floor space for light breeds; two to 2½ square feet for brown egg birds.

Ordinarily, for a small house, a shed-type roof is most economical to build. Plans for houses and equipment are shown in Chapter 11. About 1 square foot of window for every 10 square feet of floor space is adequate. The roof must be insulated; the side walls, too, preferably. A temperature of 55° F. is best for egg production. Frozen water in winter is

one of the biggest problems with small flocks. Laying birds normally drink about twice as many pounds of water as pounds of feed consumed. Without ample water they may go into a partial or full molt and stop laying. Electric heating cable and other electrical devices will keep waterers from freezing.

Cover the floor with several inches of absorbent litter—wood shavings or sawdust are most often used. Build the litter up to a depth of six to eight inches before cold weather. This makes it easier to keep dry. By removing only the wet spots, stirring occasionally, and ventilating properly, you can keep the litter drier and cleaner.

Equipment you will need

Provide a nest for every four or five hens. Nests for Leghorns should be 12 inches in all dimensions. Nests for other breeds should be 14 inches wide and high but only 12 inches deep. Place a five-inch board on edge in front of the nest to hold in a deep layer of shavings or straw. Place a pole or six-inch board in front for the birds to walk on and inspect the nests. If egg eating becomes a problem, try a community nest. A community nest is a long, dark box 15 inches wide and 4 to 6 feet long with the access holes only on the ends. The center is filled with six inches of straw or sawdust and the birds sit all together inside. The top is hinged for easy access to collect the eggs.

Locate roosts at the rear of the house away from windows. Place them two to 2½ feet above the floor to form a dropping pit. Perches are usually made of 2- by 2-inch lumber spaced 14 inches apart. Allow 10 inches of perch space on the roost. Locate the perches on a frame covered with

heavy wire netting. This keeps the birds out of the manure. The litter remains cleaner and drier. Roost frames should be movable to permit easy cleanout of the accumulated droppings in the dropping pit.

A dropping pit is designed to hold droppings for several months. For ease in handling, the floor over the pit may be made into six- by six-foot panels. There is a real saving in labor from the use of pits. However, pits harbor rats and are a breeding place for flies.

Feeders and waterers often are placed over pits.

Plenty of feeders

Allow two hanging feeders—each 15 inches in diameter with a 35- to 50-pound capacity—for 25 to 30 medium-weight layers. If hanging feeders are not available, provide at least 10 feet of feeder space or two 5-foot trough feeders with both sides open for the family flock.

Automatic feeders vary widely in capacity, but probably are more than you need. Consult the manufacturer's literature or a specialist before installing such equipment.

All feeders should be placed within 10 feet of a waterer.

Hoppers are needed for insoluble grit and calcium supplements, if these are not included in the feed. For each 100 hens, provide a 12-inch granite grit hopper box and a 12-inch hopper for oystershell or limestone grit.

Watering your pullets

Making eggs takes lots of water. The kind of waterer you use is not as important as keeping water available. You can purchase a fountain unit

from a catalog or feed supply store. You fill a tank that has an air-tight cover. As birds drink from a trough more flows in.

Water also can be provided by hand or automatically. Automatic fountains can be hung from the ceiling. A pan or pail on a small, wire-covered platform is satisfactory. The platform prevents shavings from getting into waterers and helps to keep litter drier around the watering stations.

Include a broody coop in your plan

The broody coop is used occasionally to break up birds that go broody (tend to go out of production and set on eggs in the nest). It is merely a slatted or wire-bottom coop suspended above the floor at least two feet. It is often hung from the ceiling. It also serves to isolate birds that must be separated from the flock.

It doesn't take long to cure the average broody hen. A few days is usually enough. If broodiness persists after a spell in solitary, it probably is best to get rid of the hen. She will be a continuing problem.

It's moving day

Pullets are easily disturbed. Handle them gently. Do not make unnecessary noises or frighten them. Because pullets become less excited when they are handled in the dark, many poultrymen move pullet flocks into laying houses at night.

Before moving pullets to the laying house, examine them individually. Remove weak, runty, and obviously sick chickens from the flock. Cull pullets that appear to be unthrifty or extremely slow maturing; they are unlikely to be profitable layers.

The good home flock host will make sure the feeders are well stocked and the waterers in operation. Check ventilation nest materials and other equipment that is important to get birds accustomed to their new home.

Care of pullets

Watch pullets after they are placed in the laying house; do not let them pile up or smother. Close nests at night. A night light will help birds find roosts.

Clean nests regularly; add new nesting material as needed to produce clean eggs. Clean waterers at least once every day and feeders whenever they need it. Stir litter and add new material frequently; remove and replace wet and "caked-over" spots, as necessary. Control lice and other parasites.

Regulate the working day

If chickens belonged to a union, they all would be on strike. For peak production they are worked 14 hours per day, 7 days a week.

Light stimulates egg production by entering the eye of the bird and stimulating the pituitary gland. The pituitary gland, in turn, releases certain hormones that cause ovulation.

The changing length of daylight is one of the most important features of light. Pullet chicks respond to light. Long days cause early maturity. Short days slow down maturity. Home flock owners may use the natural day length; thus early spring-hatched chicks mature in 20 weeks while summer-hatched chicks mature in 24 weeks.

It is advisable with spring-hatched pullets to continue maximum day length (16 hours) as soon as they start to lay in June or July to prevent decreasing light from hindering production. With summer-hatched pullets, do not use lights until 20 weeks of age, at which time start birds at 12 hours and increase 15 minutes per week until a 14-hour day is reached.

Artificial light is a must in the laying house. For maximum egg production, you need a 14-hour day during the fall and winter months. Provide one 60-watt bulb for every 200 square feet of floor space. Use smaller bulbs in houses having less floor area.

Lights can be turned on in the morning, in the evening, or both morning and evening to provide a 14-hour total light period. But it is simpler to use morning light. This avoids the need for a light dimmer to induce birds to roost before "lights out" at night. An automatic time clock can be used to control the lights.

In most of the United States, natural light drops below 14 hours during the first week in August. It does not return until the first week in May.

Set a good table for your layers

It takes a quality balanced ration to keep layers in shape to be high producers. We recommend that your basic ration be a mixed feed purchased at a poultry ıeed store. Laying hens need a mixture with a 15 percent protein level. Vitamins and minerals usually are blended into the commercial feed to round out the diet of your birds.

Use a good laying ration and keep it in front of the birds at all times. Feed is the biggest expense of egg production, running at about 60 percent of total cost. To prevent waste don't fill the hoppers more than one-half full. Commercial poultry rations normally contain enough calcium (3.0 to 3.5 percent) so that oystershell or other calcium supplements are not needed. No grit is necessary with present-day laying rations.

If grain is low-priced, you may want to use it to cut the cost of purchased feed. However, feeding too much grain will make your hens overly fat. When a complete 15 percent protein laying feed is used, do not feed more than one-half pound of grain per 10 hens daily. A 20 to 22 percent protein laying feed can be used with grain fed free-choice in separate feeders or spread on the ground (1½ pounds of grain for every 10 hens daily). Supplementing the complete ration with grain is most economical when low cost local grain is available.

Feeding whole grain by spreading it on the litter induces hens to scratch in the litter and maintain it in good condition.

The form of the mash makes little difference. Pellets often are offered for laying rations. Crumbles are another form frequently used for younger birds. These may cost a little more than mash but have only a small advantage. They may reduce waste or wind loss, are less dusty, and will not separate during handling.

Table scraps, garden products, and surplus milk can be useful feed supplements to reduce costs. Feeding should be limited to amounts that your birds will eat in 10 to 20 minutes. Peelings, stale bread, and leafy vegetables such as cabbage, cauliflower, or turnips, are useful. Avoid strong materials such as onions unless you relish onion-flavored eggs. Don't feed spoiled or moldy feeds or foods. Fresh or sour milk is a valuable feed. Put it in plastic, glass, or enamel containers, as the lactic acid formed will rust galvanized containers.

If chickens are fed whole grain or green forage, they should also receive insoluble grit. Grit is available in chick or hen size. Continuous feeding is not necessary, but grit should be available free choice, two or three days per month. Fine gravel is an acceptable substitute for purchased grit.

Laying hens require large amounts of calcium for eggshells. An effective way to provide it is by free choice feeding of oystershell or calcium grit. Also, eggshells can be saved, washed, dried, crushed, and fed back to the hens. Wet shells should not be fed because there is a danger of bacterial growth on the residual albumen. There also is the risk of induced egg eating.

Laying mashes containing 2 ½ to 3 ½ percent calcium supply enough calcium, *if they constitute the entire ration* (no pasture or grain). Growing chickens require only about 1.2 percent calcium in their feed. If you use the higher calcium laying feeds for growing chickens, kidney damage can result.

Feed loses its quality when stored too long. It is a good idea to buy a supply that will be used up in two or three weeks. This is particularly necessary in warm weather.

A 25-pound bag of feed should last 10 hens about 10 days, if waste is controlled and the feed is a good high-energy ration. Expect to use 80 to 90 pounds of feed per layer kept for a year.

Wild birds and rodents are competitors for poultry feed. That's why it is important to make sure the house is tightly built and windows have small opening screens. As recommended for younger birds, a steel galvanized garbage can with a tight cover offers good pest-proof storage for your family flock feed supply.

Home-mixed feed will do

Home mixing of poultry feed for small flocks is discouraged. Your hand mixed blend may not equal commercial feed in quality and it is usually easier and less expensive to buy feed from feed stores or mills. Large mills have lower production costs due to larger volume purchases of ingredients and efficient milling and mixing facilities. On the other hand, sacking and retailing costs are high, so those who have access to home-grown feedstuffs may be able to save money by home mixing. You may prefer your own mix regardless of cost. To make a cost comparison, calculate the total ingredient cost of home-mixed feed. Be sure to add the value of any home-grown grains used. (Don't overlook the alternative of feeding grain with a high-protein laying feed.)

General formulas for home mixes

	lbs. per 100 lbs. of mix		
	Starter	**Grower**	**Layer**
Coarsely ground grain (corn, milo, barley, oats, wheat, rice, etc.)	46	50	53.5
Wheat bran, mill feed, rice bran, milling by-products, etc.	10	18	17
Soybean meal, peanut meal, cottonseed meal, sunflower meal, safflower meal, sesame meal, etc. (Soybean meal is the preferred protein source. Cottonseed meal should be egg-tested type low in gossypol)	29.5	16.5	15
Meat meal, fish meal (if meat meal or fish meal is unavailable, soybean meal may be substituted)	5	5	3

	Starter	Grower	Layer
Alfalfa meal (can be eliminated if fresh pasture is available)	4	4	4
Yeast, milk powder (can be eliminated if the vitamin supplement is properly balanced)	2	2	2
Vitamin supplement (must supply 200,000 I.C.U. vitamin A, 80,000 I.C.U. vitamin D_3, 100 mg. riboflavin)	+	+	+
Salt with trace minerals (trace minerals salt or iodized salt supplemented with ½ oz. of manganese sulfate and ½ oz. of zinc oxide)	0.5	0.5	0.5
Bone meal, defluorinated dicalcium phosphate	2	2	2
Ground limestone, marble, oystershells (oystershells and grit should be fed free choice to layers)	1	2	3

*Use a combination of ingredients in each category, if possible.

A natural diet for laying hens

Some home poultry flock owners prefer a natural diet for their hens. They want to avoid some of the additives found in commercially mixed feeds. The University of Maine has developed the following formula for 100 pounds of feed that can be used with natural ingredients. It contains 16 percent protein, which is considered a good level for the laying flock.

The birds must receive direct sunlight to enable them to synthesize vitamin D. Unfortified cod liver oil can be fed in place of sunlight to supply vitamin D. The amount of cod liver oil depends on the potency of the oil. The need is for 1,000 I.C.U. per pound of feed.

Ingredient	Amount (lbs.)
Yellow corn meal	60.00
Wheat middlings	15.00
Soybean meal (dehulled)	8.00
Maine herring meal (65%)	3.75
Meat and bone meal (47%)	1.00
Skim milk, dried	3.00
Alfalfa leaf meal (20%)	2.50
Iodized salt	0.40
Limestone, ground (38% calcium)	6.35
Totals	100.00 lbs.

Calculated Analysis	Amt. (lbs.) 100 lbs.	Recommended (N.E.C.C.) (Per pound)
Metabolizable energy Cal./lb.	1252	1292
Protein	16.07	16
Lysine	0.79	0.74
Methionine	0.31	0.29
Methionine and cystine	0.55	0.54
Fat	3.67	3.33
Fiber	3.15	2.51
Calcium	2.77	2.75
Total Phosphorus	0.53	0.50
Available Phosphorus	0.44	0.42
Vitamin A activity (U.S.P. units/lb.)	5112	5290
Vitamin D (I.C.U.)	*	1000
Riboflavin (mg.)	1.36	1.38
Pantothenic acid (mg.)	3.89	4.05
Choline Chloride (mg.)	411	500
Niacin (mg.)	17.46	16.96

Definitions of Abbreviations

U.S.P.	United States Pharmacopoeia
I.C.U.	International Chick Units
Mg.(s)	Milligrams
Cal./lb.	Calories per pound
N.E.C.C.	New England College Conference

Hens need daily care

It doesn't take long for an empty waterer or feeder to throw a flock off of its production schedule. Make sure someone in the family is prepared to assume the responsibility to keep the hens supplied with the feed and water they must have to run their egg assembly line.

Your goal should be to have the flock clean up all of its feed every day—but don't cut them short. Water is especially critical. It is better to have too many waterers than to run short on a hot day.

The basic feed should always be a balanced complete ration. Don't try to get by on table scraps. Most table waste has no merit for poultry feed. Too many types are unsatisfactory because of off flavors or spoilage.

Give your hens good care every day and you'll be surprised at the number of eggs they'll provide as your reward.

Chapter 7

The hen has a marvelous assembly line

EGGS ARE PRODUCED ASSEMBLY LINE fashion within the hen's body. The yolk starts at the top of a two-foot-long oviduct. As it moves along over a 24-hour period, the white, shell membrane, and shell are added.

As many as 4,000 tiny ova are present in a hen's ovary. The full-sized yolks develop as the hen matures. Each yolk or ovum is contained in a thin membranous follicle. Blood vessels in the follicle carry nutrients to the developing yolks. When a yolk matures, the follicle ruptures along a line relatively free from blood vessels known as the stigma, and the yolk is released. The yolk is kept intact by the membrane surrounding it.

Upon release from the follicle, the yolk drops into the body cavity. There the infundibulum, or funnel, engulfs the yolk and starts it on its way down the oviduct. The oviduct is more than two feet long and is lined with glands that secrete the materials for the albumen, shell membranes, and shell. Twenty-four hours or more are required from the time the yolk is released until the completed egg is laid.

Gathering and care of eggs

Gathering a supply of fresh eggs every day is your reward for months of work in rearing a home poultry flock. Here are eggs you know are fresh and top quality.

Eggs need to be gathered at least twice each day. Your birds will "sing" as they go about their egg-laying chores in the morning hours. It's a pleasant, happy sound.

More eggs will be laid in the morning than in the afternoon. A hen will lay her egg in a nest that has been used previously by another hen. In fact, they seem to prefer to use a nest with eggs already in it.

Occasionally, a hen will want to sit on eggs after they have been laid. Just reach under her and gather the eggs anyway. You may get an occasional peck with her beak, but hens aren't very fierce.

The quality of your eggs shows how good a job you are doing with the flock. The shell should be strong, regular, and clean; the white (or albumen), thick, clean, and firm; the yolk, light colored, well centered, and free from blood and meat spots.

Egg characteristics are inherited. Bred-to-lay birds have been produced through careful breeding programs that developed desirable egg traits as well as sustained laying ability.

Feed also affects egg quality.

Lack of calcium causes eggs to have soft or thin shells. Other reasons for poor shells are insufficient minerals, lack of vitamin D, high temperatures in the laying house, and diseases. Faulty shells also may be the result of inherited factors.

Egg cleanliness begins with clean litter and nesting material. Soiled eggs should be separated from clean eggs during collection and cleaned as soon as possible. Eggs should be gathered regularly and refrigerated promptly. Leaving them in the nest increases chances of soiling and breakage.

Handle eggs carefully. Keep hands and equipment clean.

Use open plastic or rubber-coated heavy wire baskets to gather and hold eggs. These permit free air circulation, so the eggs cool in about half the time that is required in closed containers.

Place baskets in an egg storage room to cool. Keep the temperature of the egg room between 45° and 55° F. and the relative humidity at about 70 percent.

Store only clean, sound-shelled eggs if they are to be kept for some time. Soiled eggs can be washed in detergent water and then used promptly. Use eggs with shell damage promptly.

Eggs will keep well in the refrigerator for quite a while. If you want to hold some for up to six months, select only clean, sound eggs, and place them first in cartons and then inside plastic bags. This method reduces loss of moisture and carbon dioxide and maintains quality. Remember, however, that eggs will pick up flavors from certain items in the refrigerator such as onions, apples, or other items with penetrating odors.

Eggs can also be frozen for longer storage. The simplest way is to break them into a bowl, mix thoroughly without beating, and freeze in quantities to be used all at once. They must be thawed completely in the refrigerator for use and used promptly when thawed. Yolks frozen alone or without mixing with the whites will gel unless sugar or salt is added. Whites can be frozen alone without treatment.

Sell only the best

If you have surplus eggs to sell to a neighbor or a store, make sure they are your best. Deliver fresh every day. Rig up a simple candler. By turning the egg carefully the candler can see the condition and size of the air cell, yolk, and white. The light reveals defects such as blood spots, blood rings, meat spots, and development of the germ spot in an egg that might be fertile. (There shouldn't be fertile eggs in a tight laying house with no roosters.)

Stock up on one-dozen egg cartons. Perhaps you can rotate them with your customers. They offer a safe way to carry the eggs and they look professional.

How eggs are graded

Egg grading entails grouping eggs into lots having similar characteristics. Sorting is based upon quality and size or weight. The advantage of grading eggs is to provide a uniform "yardstick" or system of measuring a product that has much variability in its size and quality.

Grades based on the U.S. Standards for Quality of Individual Shell Eggs are determined by candling. The "Standards" list specifications for the following qualities:

Quality ratings

AA Quality—The shell must clean, unbroken, and practically normal. The air cell must not exceed one-eighth inch in depth, may show unlimited movement, and may be free or bubbly. The white must be clear and firm so that the yolk is only slightly defined when the egg is twirled before the candling light. The yolk must be practically free from apparent defects.

A Quality—The shell must be clean, unbroken, and practically normal. The air cell must not exceed three-sixteenths inch in depth, may show unlimited movement, and may be free or bubbly. The white must be clear and at least reasonably firm so that the yolk outline is only fairly well defined when the egg is twirled before the candling light. The yolk must be practically free from apparent defects.

B Quality—The shell must be unbroken, may be abnormal, and may have slightly stained areas. Moderately stained areas are permitted if they do not cover more than one-thirty-second of the shell surface if localized, or one-sixteenth of the shell surface if scattered. Eggs having shells with prominent stains or adhering dirt are not permitted. The air cell may be over three-sixteenths inch in depth, may show unlimited movement, and may be free or bubbly. The white may be weak and watery so that the yolk outline is plainly visible when the egg is twirled before the candling light. The yolk may appear dark, enlarged, and flattened, and may show clearly visible germ development but no blood due to such development. It may show other serious defects that do not render the egg inedible. Small blood spots or meat spots (aggregating not more than one-eighth inch in diameter) may be present.

Dirty—An individual egg that has an unbroken shell with adhering dirt or foreign material, prominent stains, or moderate stains covering more than one-thirty-second of the shell surface if localized, or one-sixteenth of the shell surface if scattered.

Check—An individual egg that has a broken shell or crack in the shell but its shell membranes are intact and its contents do not leak.

Leaker—An individual egg that has a crack or break in the shell and shell membranes to the extent that the egg contents are exuding or free to exude through the shell.

Weight classes of U.S. consumer grades for shell eggs

Size or weight class	Minimum net weight per dozen
Jumbo	30 oz.
Extra large	27 oz.
Large	24 oz.
Medium	21 oz.
Small	18 oz.
Pee Wee	15 oz.

Are your hens egg eaters?

Laying hens that eat eggs pose a costly, stubborn problem for poultrymen. Egg eaters, once they have acquired the habit, are difficult to stop; and they will eat both good and bad shell eggs. Even worse, the vice spreads quickly from one hen to another, complicating the problem.

Both caged and floor layers and large and small operations are vulnerable to the chicken's vice, although it is more difficult to break the habit with caged layers, since fewer corrective measures are available.

Empty eggshells, either on the floor or in nests or cages, are indicators of the problem.

Breaking the habit: The best cure for egg eating is to prevent it. Egg eating usually begins with a few birds that eat eggs that have been broken. Therefore, those things which cause egg breakage should be eliminated or reduced. Common causes of egg breakage include: inadequate nesting space, insufficient nesting material, failure to gather eggs frequently, and the laying of large numbers of soft or thin shell eggs.

Correcting the problem with floor birds: Provide plenty of nests, about one for each four or five hens. Use plenty of nesting material and replenish often. Placing a section of cut-to-fit cardboard in the bottom of the nest is helpful when nesting material has been pushed aside. Darkening the nests may also help. Using community nests usually results in less breakage and egg eating. Remove the broody hens so that fighting will be reduced and fewer eggs broken. Placing glass eggs or torn pieces of white paper on the floor will often frustrate the birds and sometimes break the habit.

How many eggs to expect

The home flock poultryman has done a good job if hens average 18 dozen eggs per year. That's 216 eggs per hen. If you have 15 good layers, you would gather an average of about 9 eggs per day. Averages are tricky when it comes to egg production. Hens tend to hit peak production at about 32 to 34 weeks of age. After that there is gradual tapering off. You might get as many as 12 eggs at a peak period. On a cold morning in February there might not be enough for an omelet. If you give 10 to 15 hens good management there will be plenty of eggs for the family all through the year with some left over to sell most of the time.

Chapter 8
When hens stop laying

CULLING THE LAYING HENS ON our farm was done by our favorite hatcheryman. Harvey was almost a partner in our poultry operation. He sold us the chicks, culled the hens, bought the culls, and hauled them away.

Harvey was a man of many talents. He could look a hen in the eye, look her in the tail end, puff on his cigar, and talk a blue streak all at the same time.

Once you catch on to the tell-tale signs of production, culling really isn't that tough. My experience in culling came in Poultry 101 at Iowa State. Our class went out to the poultry farm on Saturday mornings to learn and practice this art. I was courting my wife at the time and descriptions of my growing ability to judge females fascinated and/or terrified her.

Sooner or later you are going to notice a drop off in the number of eggs you gather each day. As the chart on page 85 of this chapter shows, hens peak at about 34 weeks of age; then there is a leveling off in production.

There are many things that can cause hens to stop laying eggs. The most common causes are decreasing day length, disease problems, advancing age, poor feeding practices, and stress.

Molting is a natural process whereby a bird renews its feathers. In chickens it is often associated with a seasonal pause in egg laying; however, a molt can occur any time if a hen encounters severe stress. A rapid feather loss by the entire flock is usually a sign that something serious has happened (such as lack of water or severe chilling).

Decreasing day length occurs naturally between June 22 and December 22. This change frequently causes hens to molt and stop laying for about two months when they receive only natural light. To prevent this, provide artificial light to maintain a constant day length of at least 16 hours per day. For small flocks it is usually most practical to provide continuous light. One 40-watt light globe mounted six to eight feet above the floor in the center of the pen will provide adequate light for pens up to 15 feet by 15 feet. The light can be turned off during daytime, but if you forget to turn it on, even once, hens may start to molt and stop laying. If you want to turn the light off during daylight, a time clock is recommended. Morning and evening lights (4 a.m. to 8 p.m.) are recommended with a slight overlapping of artificial and natural light in the morning and evening.

Disease problems can occur under the best of conditions. Often the first sign is a drop in egg production. Other symptoms you might see include molting, dull and listless appearance, coughing, lameness, and death. You can't tell what the problem is just by looking at the flock. Even skilled poultry veterinarians must make a careful post-mortem examination to accurately diagnose poultry disease problems. Remember that a few losses must be expected; if you see one sick bird, isolate or sacrifice it and watch the rest of the flock carefully. If other birds show similar symptoms, it is advisable to seek help from a veterinarian.

Cull the freeloaders

A hen eats about seven pounds of feed a month whether she is laying eggs or not. You'll want to get rid of these freeloaders as soon as possible. Weeding out the poor performers in the hen house is called culling.

Culling is particularly valuable if you keep hens for a second year of egg production. If hens are properly culled, the ones remaining at the end of the first laying season are the highest producers. The healthy, vigorous, fast-molting hens in this group will be the most desirable second-year layers.

Whenever you locate unproductive birds in the laying flock, cull them. When you cull during the period of peak production (generally the first six months after production begins), cull by flock inspection to remove all diseased, injured, or unthrifty birds.

When you cull after the peak period of production, cull by individual inspection to identify and remove unthrifty birds and low producers.

It is a good idea to cull at night. Culling when it is dark is less likely to frighten hens than culling when there is light. Use a flashlight to identify poor layers on the roost at night. A flashlight that produces a blue light is preferable.

Egg-laying indicators

Good layers will have large, soft, red combs and wattles and bright, prominent eyes. The good layer will have a good body size, her vent will be enlarged and moist, and her pubic bones will be spread apart and free from fat. Yellow pigmentation in the vent, eye ring, ear lobe, beak, and shank will disappear in good layers.

Nonlayers will have shrunken, pale, hard combs and wattles. Their body size will usually be small, their vents will be small and dry, and the pubic bones will be close together and covered with fat. When a hen stops laying, yellow pigmentation will return to the vent, eye ring, ear lobe, beak, and shank.

Here is a more detailed description of things to look for when you cull:

Comb and wattles

When a hen is not laying, her comb and wattles are dull, dry, and shriveled.

A laying hen has a large, smooth, bright-red comb and full, smooth wattles. When laying ceases, the comb and wattles shrivel again.

Vent

The nonlayer's vent is shrunken, puckered, dry, and round. It has a yellowish color. As a pullet begins to lay, the vent enlarges and becomes oval-shaped. A good layer has a smooth, moist, almost white vent.

Pubic bones

The two small bones at the sides of the vent are the pubic bones. In nonlayers pubic bones are thick and stiff, because fat accumulates around the bones during nonproduction. The pubic bones are close together when a hen is not laying. The distance between them is one finger width or less (a finger width—or finger—equals three-quarters inch).

These bones spread apart when egg laying begins. Fat gradually disappears and the pubic bones become thin and pliable during continuous production. In a layer, pubic bones are at least two finger widths apart.

Abdomen

The size of the abdomen, except in an excessively fat hen, is a good indication of laying.

The abdomen of a nonlayer is hard and contracted: the skin feels coarse and thick. The abdomen of a layer feels soft and pliable.

Pullets and nonlaying hens have a depth of about two finger widths between the pubic bones and the keel. Laying hens have a depth of about four fingerwidths.

Bleaching of pigment

As birds produce eggs over a long period, their bodies bleach, or lose yellow pigment from skin and beak.

Bleaching is readily seen in yellow-skinned breeds and in birds that are fed a yellow-pigmented diet. In white-skinned birds or those fed a diet low in yellow pigment, bleaching is difficult to detect.

Hens that have lost pigment are usually—but not always—high producers. Unthrifty and diseased birds that are not laying sometimes may be bleached. Therefore, bleaching must be considered in relation to other culling guides in selecting unproductive birds.

Bleaching of a hen's body begins at the same time egg production begins. The yellow pigment is diverted from the skin to egg yolks. The amount of bleaching is determined by the length of time the supply of pigment has been diverted.

The different parts of the hen's body bleach in regular order—the vent, the eye ring, the ear lobe, the beak, and the shanks.

Vent changes

The first and most rapid change occurs in the color of the skin around the edges of the vent. When a yellow-skinned pullet begins to lay, the color fades from the vent and disappears in a few days.

A yellow vent shows that a hen is not laying, and a white, pink, or bluish-white vent indicates that a hen is laying. The bluish-white vent indicates long-term production.

Eye ring and ear lobe

The eye ring starts to bleach soon after the vent. Usually it is completely bleached in two to three weeks after the hen begins laying.

The ear lobe loses its yellow pigment in three to four weeks. In Leghorn varieties, a bleached ear lobe indicates longer production than a bleached vent or eye ring.

Beak

The color of the beak fades from the base to the tip. In most high-producing breeds, a nonlayer's beak is yellow. As a bird begins to lay, her beak loses color from the base. It takes about six weeks of continuous egg production for a beak to bleach almost white.

The lower beak bleaches more rapidly than the upper. Bleaching of the lower beak can be used as a basis for culling when the upper beak contains dark brown or black pigment. This dark pigment is frequent in Barred Plymouth Rocks and Rhode Island Reds.

Shank coloring

Bleaching of the shanks is a good indication of long-time production, because it occurs more slowly than bleaching of the beak.

The color recedes first from the lower edge of the scales on the front of the shank. The rear of the shank gradually bleaches with continued production until little pigment is left except in the scales of the hock joint. Continuous egg production for two to five months is required to bleach the shanks completely.

Pigment returns

When a hen stops laying, the yellow color comes back. The speed of return depends on the kind of feed she gets; large quantities of green feed and yellow corn in the mash-and-grain ration will restore yellow color quickly. Pigment returns to the hen's body in the same order as it bleaches—in the vent first and the shanks last.

In the beak, the color first reappears at the base. As the hen remains out of production, the returning pigment moves across the beak until it reaches the tip and the whole beak is again a normal yellow.

During long periods of broodiness when a hen is not laying, the color returns to the skin and beak.

When to cull hens

Chickens lay best during their first production year, starting at about 20 weeks of age. They can be expected to reach a peak production varying from 60 to 90 percent at about 34 weeks of age. Following this peak the rate of egg laying will decline until a molt occurs. Commercial flocks usually are processed for meat after 14 to 20 months of egg production, unless they are force molted.

Winter production pauses accompanied by a molt are common in small flocks. They can be prevented to some extent by providing proper lighting, adequate protection from cold weather, and plenty of feed and water. Hold off on culling if your hens react poorly in tough weather.

Broodiness is another trait that merits culling. When a hen persists in sitting on eggs she is doing what comes naturally. But it isn't what you want in a laying house. Try to break up broodiness by putting the hen in a wire-floored cage for a week or so. If that doesn't cure her inclinations to raise a family, mark her for culling.

Quick guide to productive hens

Character	Layer	Nonlayer
Comb	Large, smooth, bright red, glossy	Dull, dry, shriveled, scaly
Face	Bright red	Yellowish tint
Vent	Large, smooth, moist	Shrunken, puckered, dry
Pubic bones	Thin, pliable, spread apart	Blunt, rigid, close together
Abdomen	Full, soft, pliable	Contracted, hard, fleshy
Skin	Soft, loose	Thick, underlaid with fat

Character	High producer (continuous laying)	Low producer (brief laying)
Vent	Bluish white	Yellow or flesh color
Eye ring	White	Yellow
Ear lobe	White	Yellow
Beak	White	Yellow
Shanks	White, flattened	Yellow, round
Plumage	Worn, soiled	Not much worn
Molting	Late, rapid	Early, slow

What to do with culls

Non-productive hens are ready for the stew-pot. If there is a market for them in your community, you may want to sell them. They can be an important part of your poultry meat program for the family. Butcher and utilize them in recipes that call for long cooking. Remember, a hen isn't a tender broiler or fryer. It will take some ingenuity in the kitchen to please your family.

Hens culled because of disease should be killed and disposed of. Incinerating is a smelly business that will bring objections from your family, the neighbors, and possibly local authorities. Deep burial is the best method.

Production Curve for Laying Hens

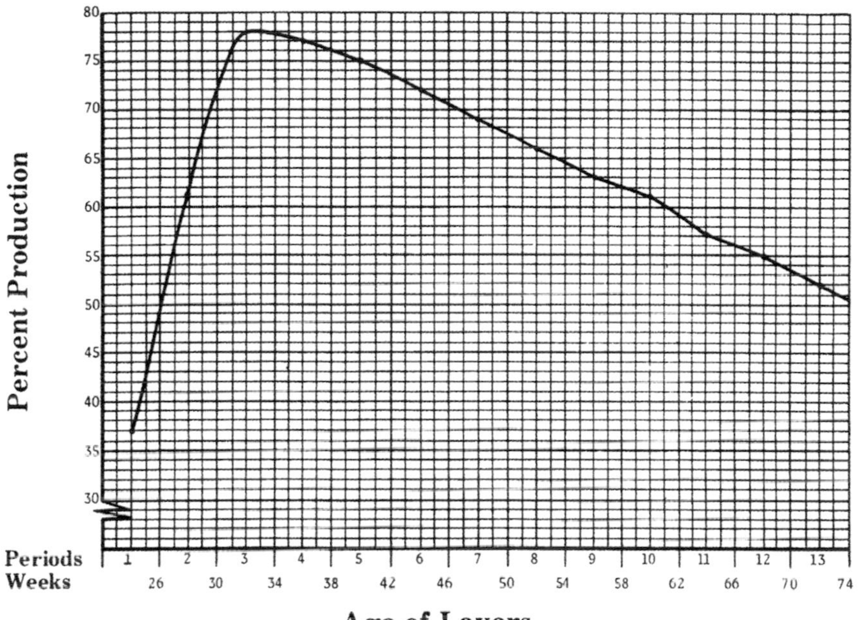

Chapter 9
Fast-growing meat birds

ALL OF THE EXPERTS ADVISE against raising Leghorn cockerels. "Aren't worth the feed," they declare. That's probably true, but this farm boy grew up on them.

We had chicken every Sunday beginning the week the first unlucky rooster weighed two pounds. One of the Sunday morning chores was to catch a couple of birds and take them to the chopping block. A stroke of the corn knife, a quick bleed, and the bird was ready for plucking. Small boys were eligible for this chore. Mother could cut up a chicken about as fast as anyone I ever saw. None of this leg-and-thigh-together business at our place. Or breast with wish bone. We cut that bird into as many pieces as possible.

After church, Mother quickly fried the bird and made the gravy. It was a great Sunday dinner. My first choice was a leg and I usually got the wishbone too. I probably was best at eating. About the time I was in high school, I was considered capable of the corn knife bit. But let it be said that my heart really was on the chicken's side.

Today's broiler chicken is a remarkable bird that can convert feed to meat more efficiently than any other creature. Most broilers you buy at the store are raised in huge plants under scientific conditions. Birds are brooded and raised to eating size in six to eight weeks. Many produce a pound of meat on only two pounds of feed. Contrast that with a pig that takes five pounds of feed to make a pound of gain or a steer that takes eight.

You, too, can produce meat birds in a remarkably short time. In about two months you can go from a downy chick to a fryer ready for the table.

A well-planned and well-managed flock can be a good source of fresh poultry meat. However, you aren't going to save much money over purchased broilers. The large scale commercial operations are so efficient that broilers are the cheapest of the meats in the supermarket. Your goal should be to produce just enough poultry meat for your family use. You may want to have some for friends and neighbors. But if you start selling them in quantity you'll run across laws that require official inspection of slaughter.

One of the problems with broilers is their uniform growth and efficiency. All of a sudden, after a couple of months of feeding the entire flock is ready for eating. Chicken every Sunday isn't bad but chicken every day is too much.

You can start eating meat birds at five to six weeks of age. This is the bird marketed as the Cornish game hen. Regular broilers are ready at seven to nine weeks of age. Grow them a little longer and at 12 to 14 weeks they make plump, tasty roasters.

This adds up to about 10 weeks of peak meat production. If you raise 20 birds for the home poultry flock, that's 2 per week. You can do some freezing to stretch the season. Basically, most families will find 20 to 25 birds is about right. If you have friends and relatives, 50 can be handled as easily as 25.

Type of chicks to buy

Birds grown for meat—capons, roasters, and broilers—usually are crosses or hybrids. Cornish varieties—Silver and Dominant White—are popular for use as the male parent because their chicks develop meaty breasts and legs. Varieties and strains with white plumage often are used for female parents because they produce birds with excellent market appearance.

Others are crosses such as Cornish and Plymouth Rock or New Hampshire. These crosses have been bred for the most economical conversion of feed to poultry meat. They feather rapidly and mature early. Cornish breeding helps them develop meaty breasts and legs.

Some breeds such as White or Barred Plymouth Rocks, Rhode Island Reds, and New Hampshires are used for home flock egg and meat birds. They usually don't grow as rapidly as the crosses and take more feed per pound of weight gained. Leghorn males do not make good meat birds and are unprofitable even if day-old chicks are given to you.

Where to get chicks

Hatcheries produce broiler chicks by the millions every month. Ask your local hatchery to give you some of the same birds he is delivering to commercial producers.

Another source is to purchase day-old cockerels of one of the larger laying breeds. If the hatchery is selling females for a laying flock, you may be able to get the cockerels at a low price. However, for efficient meat production, you can't beat the bird bred specially for the broiler market.

Hatcheries, breeders, and dealers are continually improving the stock they sell for meat production. Check the growth rate, feed conversion, and market acceptance records of available stocks before ordering meat-type chicks.

For meat production, almost all flocks are started from day-old chicks. Birds usually are purchased as straight-run chicks—about half pullets and half cockerels.

Brooding the chicks

Starting broiler chicks is similar to those for a laying flock. Review the housing, heating, feeding, and watering suggestions for young chicks detailed in Chapter 4.

Since these birds grow so quickly, more space on the floor and at feeders and waterers is needed almost on a weekly basis.

Place feed on chick box lids or trays from cut-down cardboard boxes for the first few days. Feed and water should be available to the chicks as soon as they arrive. Chicks need one lineal inch of feeding space for the first two weeks, two lineal inches for two to six weeks, and three lineal inches after six weeks. Feed wastage is minimized by filling hoppers half full and adjusting feeder height or size to birds' size.

Provide a one-half-gallon water fountain for 25 chicks the first 2 weeks. Increase the number or size of waterers from 2 to 10 weeks to provide 10 inches of watering space per 25 birds or one-gallon capacity for 15 chicks. Provide one-gallon capacity per 10 birds for older birds if using fountains or one inch of trough space.

Housing needs

Broilers, capons, and roasters may be raised either in confinement or on range. The tendency is to raise them in confinement. Brooder houses are described in Chapter 11.

Floor space per bird should be increased to one square foot from 6 to 10 weeks. From 10 weeks on they will need additional space—at least 2 to 3 square feet if they do not have access to a yard or range. Butchering some of the birds at various ages increases floor space for remaining birds.

Most broiler growers provide lights 24 hours a day as long as birds are on the farm. A usual lighting system allows one 60-watt lamp for each 200 square feet of floor space, or about ¼ watt per square foot.

Maintain litter in good condition, removing wet spots and caked litter as necessary to keep the floor dry and birds comfortable. Some producers reuse litter for two to three broods of broilers if no disease or parasite problems have been encountered.

Feeding broiler chickens

Broilers need a high-powered ration to make the gains they are bred to produce. Your local feed dealer can supply you with a ready-mixed feed that is just right for your birds. It really doesn't pay to fiddle around mixing your own feed for a 25-bird home poultry flock.

Broilers need a feed that contains 20 to 24 percent of protein for the first six weeks. The exact percentage of protein needed depends on the feed's energy content.

All broiler rations should be fed without grain. A sample starter ration for broilers is listed in the following table. Chick-sized granite grit may be fed in boxes or scattered on mash.

When using commercial feed, follow the manufacturer's directions.

At about six weeks, place broilers on a finishing mash that has an increased energy level and reduced protein level. A sample finishing mash also is listed in the table. Pellets often are used for broilers from six weeks until they reach market weight.

Give capons and roasters the kind and amount of feed recommended for broilers during the first six weeks. After changing to finishing mash, supply cracked corn to roasters and capons in the afternoon. Gradually increase the grain until birds are getting equal amounts of corn, mash, and pellets at 12 weeks of age. Remember, fast-growing broilers must have a balanced diet including both grain and protein. Match ration to growth of birds.

Formula for home-mixed feed

If you decide to mix your own feed, here is a good formula for both starter and finisher ration. This also can be a guide to selection of feed at your local store. Compare these ingredients with those listed on the bag tag.

Complete Broiler Feeds		
Ingredient[1]	**Starter Percent**	**Finisher Percent**
Ground yellow corn	59.40	66.61
Fish meal (60 percent)	6.00	5.00
Poultry by-product meal	5.00	2.50
Corn gluten meal	4.00	
Soybean oil meal (44 percent protein)	18.00	16.50
Dried whey		1.50
Alfalfa meal (17 percent protein)[2]	2.00	2.00
Dried distillers' solubles	2.50	2.00
Calcium carbonate	1.25	1.00
Dicalcium phosphate (18 percent phosphorus)	.60	
Bonemeal, steamed		1.50
Salt, iodized	.30	.30
Manganese sulfate (65 percent grade)	.05	.05
Vitamin A supplement (4,000 USP units per gram)	.05	.05
Vitamin D_3 supplement (1,500 ICU per gram)	.06	.06
Vitamin B_{12} supplement (12 milligrams per pound)	.05	.05
Riboflavin supplement (227 milligrams per pound)	.50	.50

[1] Should also include a coccidiostat at level recommended by manufacturer.

[2] Meal with guaranteed vitamin A content of 100,000 International Units per pound.

Ingredient	Starter Percent	Finisher Percent
Choline supplement (25 percent grade)[3]	.10	.10
DL-methionine (feed grade)	.04	.18
Antibiotic supplement (10 grams per pound)[4]	.05	.05
Arsonic acid (10 percent)[5]	.05	.05
Total	100.00	100.00
	Grams per ton	
Niacin	25	25
Calcium pantothenate	5	5
Vitamin E	5	5
Vitamin K	1	1

[3] Contains 25 percent choline chloride.

[4] Follow the recommendations of the manufacturer.

[5] Contains 10 percent 3-nitro, 4 hydroxyphenylarsonic acid. Other compounds, including sodium arsanilate or arsanilic acid, may be used at a level recommended by manufacturer. All arsenical compounds to be used in poultry feeds are subject to approval by Food and Drug Administration, U.S. Department of Health, Education and Welfare.

Amount of feed required

This chart shows the amount of feed you should expect to give birds up to slaughter age. It is set up for 100 birds. Apply the proper percentage to the number of birds you plan to raise.

The second chart shows the amount of water required per 100 birds. Plenty of water is very important to both growth and bird health. Remember that chickens need more water on hot summer days. Consumption may be double that in moderate weather.

Feed Consumption in Pounds/100 Birds					
Age in Weeks	Avg.Wgt.	Feed Conversion	Daily	Weekly	Cumulative
1	0.23	0.70	2.3	16	16
2	0.47	1.11	5.0	35	51
3	0.82	1.37	8.5	60	111
4	1.23	1.58	11.7	82	193
5	1.72	1.72	14.4	101	294
6	2.29	1.85	18.3	128	423
7	2.92	1.94	20.5	144	567
8	3.52	2.08	23.8	167	734
9	4.13	2.20	25.4	178	912

Water Consumption in Gallons/100 Birds			
Age in Weeks	Daily	Weekly	Cumulative
1	0.5	3.5	3.5
2	1.2	8.4	11.9
3	2.0	14.4	26.3
4	2.8	19.6	45.9
5	3.5	24.5	70.4
6	4.4	30.8	101.2
7	4.9	34.3	135.5
8	5.7	39.9	175.4
9	6.1	42.7	218.1

How to raise capons

A capon is any male chicken that has been castrated. Following this surgical operation, the bird fattens more readily and produces tenderer meat. Capons sell for a higher price per pound than broilers or roasters, because they require more labor and cost more to produce. They do not use feed as efficiently as lightweight birds.

Most poultrymen caponize when cockerels are three to five weeks old. The operation is of little value after chickens are two months old.

Caponizing is a surgical operation that no one should try without training. If you want to raise capons find someone who can do the operation for you. Or if you are really serious about it, learn the technique from a good instructor. Testicles are removed through incisions between the ribs of the bird's body. Although the operation takes some skill, large numbers of capons are produced each year. There are few problems when it is done right.

After the operation, put capons in clean quarters, separated from other chickens for two or three days. Capons do not need special care after they recover from the initial shock of the operation.

Capons usually are raised to five to six months for best weight and finish. Capons must be kept to this age for them to make their more rapid gains in weight over cockerels of the same age.

Capons can be started on the same feeding and management program as other meat birds. During the growing period (7 to 12 weeks) it may be advantageous to provide a lower energy feed with about 17 percent protein so that the birds develop more fully before reaching the heavier weights. Birds fed for too rapid a gain may be more prone to leg weakness and breast blister problems.

After 13 weeks capons can be placed on a higher-energy 15 to 16 percent protein finishing feed, which will push them for maximum weight gain. Cracked corn can be supplied to capons and roasters in the afternoon after changing from the starting mash. Gradually increase the grain until birds are getting equal amounts of corn and broiler finisher mash at 12 to 15 weeks. Feed a small amount of grit once a week whenever birds are fed corn.

Capons are in greatest demand at Thanksgiving and Christmas. However, there is a growing demand for capons during other holiday seasons, especially the Jewish holidays in September. Because capons take five to six months to grow and finish properly, the best time to caponize for these markets varies from spring to early fall.

Avoiding problems

Breast blisters are sometimes a problem in roaster and capon production. They are caused by contact of the keel bone with the litter and poultry equipment and the irritation results. Maintaining good litter condition, preventing overcrowding, using types of equipment without sharp edges, and feeding programs that develop good body structure before heavy weights are reached are all factors that can help reduce incidence of breast blisters.

Cannibalism and feather picking are other problems that may develop with poultry meat birds. Debeaking at the hatchery will eliminate these problems. Various factors such as crowding, nutrient deficiencies, inadequate ventilation, too little drinking and eating space, too much light, idleness, and the appearance of blood on an injured bird contribute to picking. Good management can frequently ward off cannibalism. In small flocks a pick-paste remedy can be used with much success if the problem is not out of hand.

Isolation from other birds is a first means of preventing disease. Restrict unnecessary traffic of people and pets to the poultry flock. If different ages of chickens are present, physically separate the flocks as much as possible and care for the younger birds first. It is easier to control diseases and parasites if birds are kept confined. Obtain chicks that are from pullorum-typhoid clean stock. Coccidiosis can usually be prevented by good sanitation and a low-level coccidiostat drug in the feed. Rotate range areas so that birds are not on the same ground year after year. Adjust ventilation to avoid moisture and ammonia build-up in the house. Clean waterers daily and periodically wash them with a sanitizing solution. A local veterinarian, county extension agent, or commercial field serviceman can assist with flock health and other management problems or direct you to competent help.

Chapter 10
How to butcher your meat birds

IF YOU ARE RAISING MEAT birds they can be home dressed with a little skill and persistence. Once you have killed and dressed a few broilers you'll be surprised how quickly and easily it can be done. Killing a chicken and dressing it is not for the faint hearted. It does take a certain amount of grit to kill a chicken, bleed it, and do the butchering work. However, the job isn't much different than cleaning a fish or a duck shot during the hunting season.

You may want to have a commercial slaughterhouse do your broilers for you. You could take the entire flock to the shop and have them commercially killed, dressed, and frozen.

Poultry can be dressed at home with little or no special equipment. The primary concern is sanitation. Use clean equipment and prevent contamination of the carcass with fecal material or the contents of the crop or intestine.

All birds to be slaughtered should be fasted (no food) for 24 to 34 hours before killing. Fasting empties the digestive tract of feed and ingested matter, thus reducing possible contamination of the carcass. To fast birds, pen them in a wire-bottom cage during the fasting period. A wire-bottom cage prevents birds from picking up feathers and litter. Provide fresh water to drink during the fasting period. Water-starved birds will be dehydrated and the skin will appear dark, dry, and scaly when the feathers are removed.

Holding pens should be of a kind that prevent the soiling of the birds' feathers. Dirty birds contaminate the scald water and on occasion scald water is drawn into the lungs and air sacs, so keep scalding water clean by changing it often.

Killing, bleeding, scalding, picking, pinning, and singeing

Killing: Suspend the bird by its feet in a shackle, by a rope, or in a killing cone. Clasp the head in one hand, pulling down for slight tension and to steady the bird. With a sharp knife, sever the jugular vein just behind the lower jaw on one or both sides. Make the cut at least two inches long and into the base of the skull.

Bleeding: To avoid excessive splattering of blood, if a killing cone is not available, hold the bird by the head until bleeding and flopping stops, which is about three minutes. The blood can be caught in a container partially filled with water for later disposal or directed into a sink with a stream of cold water to hemolyze the blood and prevent clotting.

Scalding: The scald temperature for broilers, roasters, and capons is from 125° to 140° F. Boiling water should be kept on a stove nearby to keep the temperature of the scald water hot enough. The hotter the water the shorter the scald time and the more chance of overscald.

Scald time is normally 60 to 90 seconds in water 125° to 140° F. Two dips of 20 to 30 seconds each may be adequate for the hotter water. Grasp the chicken by its feet. Immerse head-first in the scald water to middle of scales on shank. Move bird up and down and from side to side while in the water to insure even and thorough scalding. Test the release of tail and wing feathers immediately afterwards. Repeat short dips of scalding until large tail and wing feathers are easily pulled.

Picking: The bird can either be held, suspended, or placed on a table for picking. Use a slight pressure with a gentle rubbing action for fast, easy, and thorough removal of feathers and pinfeathers. This should be done as rapidly as possible. A suggested picking sequence is wings, tail, legs, breast, neck, and back, in that order. Establish the sequence that works best for you. Five minutes is adequate for an experienced feather picker if the bird has been scalded properly.

Pinning: Pinning is easier and faster if it can be done under a gentle stream of cold water. Again use slight pressure and rubbing motion. A pinning knife or any other dull instrument will be helpful in getting the few hard-to-get pinfeathers. Use pressure starting below the follicle to squeeze out the pinfeathers. A few may have to be pulled.

Singeing: Semi-mature and mature chickens and turkeys have a few hairs that are seen when feathers are removed. Singe these hairs by slowly rotating the defeathered bird in an open flame. Singe torches are available commercially, but a bottle gas torch or open flame on a gas range work very well. Do not burn yourself or start a fire.

Evisceration of the bird

Remove feet at hock joint: Use a boning knife or shears. The bird can be in a shackle or on a table. Hold feet with one hand so as to put backward and upward pressure on hock joint. With sharp knife, cut through hock starting on inside joint surface. Hold shank and pull joint into knife to aid in cutting through the joint. Slight movement of the bird's feet may help to complete the cut.

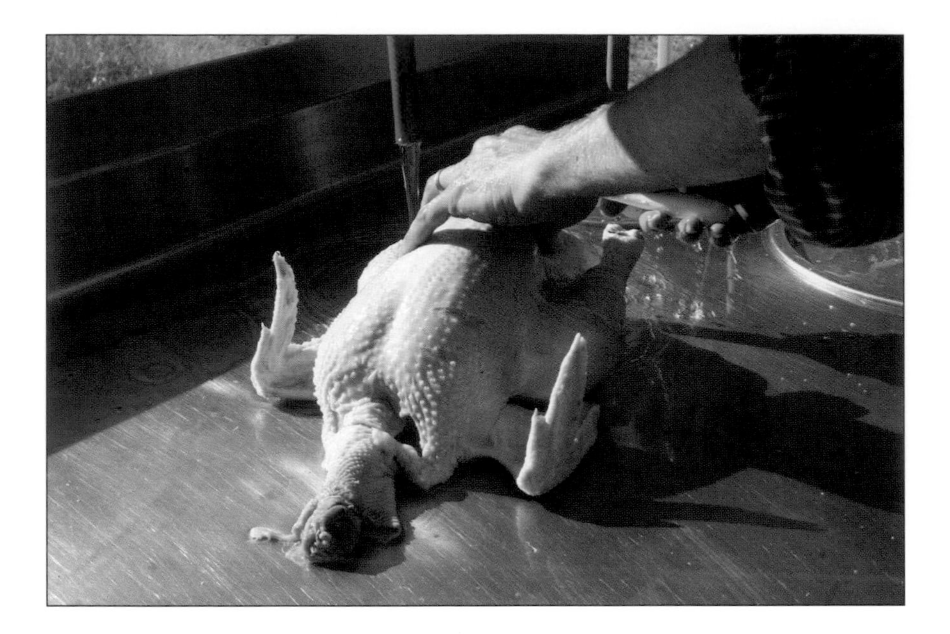

Remove oil gland: Bird still suspended or on table. Start the cut one inch forward of the oil gland nipple, cut deep to the tail vertebra, then follow the vertebra to end of tail in a scooping motion.

Remove head: Cut off the head between the head and the first neck vertebra with a knife or shears.

Split neck skin: Insert knife through skin at the point of shoulders, then cut forward guiding the knife up the back of the neck.

Pull crop and trachea (windpipe): Pull skin loose from neck. Pull out crop and trachea. This can be done after removing lungs.

Remove neck: Neck can be cut off at this point or after chilling. Cut neck muscle with knife or shears, then remove neck by twisting. Remove neck flush with body junction. Wash and chill neck.

Abdominal openings

There are two kinds of cuts used for the abdominal opening: The midline cut, which is a vertical cut from the keel down to the vent; and the bar cut, which leaves a horizontal strap of skin across the abdomen. This can be used to restrain the legs when dressing is completed.

Midline cut: Gently pull abdominal skin and wall forward and up away from entrails (viscera), then make cut through the skin and wall starting with the knifepoint just to the right of the point of the keel (over gizzard) and extend cut to the tail alongside the vent. Make cut slowly and do not cut intestine. This can be done by not cutting deeply into the abdominal cavity or by holding the wall up and away from intestines. Don't make deep, fast cuts. Complete the cut around the vent. The vent cut can be done easily and safely by keeping the knife next to the back and tail, as far as possible from the vent. This cut is routinely used on broilers and small roasters. Trussing materials are needed if capons or turkeys are opened in this manner.

Bar cut: This cut is especially useful for large birds such as turkeys and capons. This procedure does not work well on extremely short-legged birds or birds with large deposits of abdominal fat. The bar cut provides a natural and simple method of trussing the carcass after processing. The procedure is done in three steps:

Step 1: Preferably with the bird suspended by the hocks, make a half circle cut around the vent. Insert a short, thin-bladed knife into abdominal cavity above vent next to inner surface of tail vertebrae. Cut laterally in each direction to pin bones or slightly farther.

Step 2: Insert index finger into opening cut, up over intestine. Using finger as a guide, extend cut with shears on around to free the vent. Gently pull cloaca and a few inches of intestine out to prevent it from dropping into cavity.

Step 3: Complete bar cut by making a horizontal cut (side to side of bird) about three inches long. This cut should be about one and one-half to two inches below the point of the keel. Below this cut there will be a "bar" of skin about one and one-half to two inches wide. Thread the cloaca and intestine over the skin bar.

Draw viscera (pull entrails): Stretch the abdominal opening, insert hand as far forward as possible, breaking attachments of organs to the wall as you go. Pick up heart between index and second finger, cup hand and gently pull all viscera out, using a slight twisting motion as the viscera is brought out of the abdominal cavity. Leave viscera hanging to bird.

Harvest giblets: With knife or shears clip off liver—avoid cutting the gall bladder. Gall bladder may be pinched off of liver. Trim liver out, rinse. Pull off heart, trim off the heart sac and the auricles (top part), and rinse. Clip gizzard attachments (stomach and intestine). Force thin point of shears through gizzard and cut thin wall side. Open under a gentle stream of water. Peel gizzard lining, rinse. Wash all giblets and chill. Break any remaining attachments of viscera and put viscera in offal container.

Remove gonads (ovaries or testes): Pull by hand, or clip attachments with shears first and then remove by hand, or remove with a lung scraper. The gonads are attached to the backbone by ligaments (tissues) just above the liver. They will be on the floor of the cavity when the bird is on its back. Only surgical capons will be without gonads. Each bird should be checked anyway.

Remove lungs: Remove with lung scraper or by hand. By hand, use index finger to break attachments of lungs. Start next to ribs and roll finger toward center. Repeat for other lung and pick out lungs. Make certain body cavity is clean.

Wash: With hose or under faucet. Thoroughly wash inside of carcass. Repeat washing procedures for outside and rub off all adhering dirt, pin-feathers, loose cuticle, blood, and singed hairs.

Chilling and packing

Prechill: Put the bird carcass in chill container filled with tap water. Either let the water overflow continuously at a slow rate or periodically change the water. This will cool the carcass to water temperature and further clean the carcass. Use only water that is safe for drinking.

Chill: Ice and water are necessary to chill processed poultry effectively. The carcass temperature should be brought down to 40° F. before packing in bags. Large capons will require three hours or longer to chill properly.

Turkeys that are to be frozen should be held in 40° F. chill water for 18 to 24 hours before packing and freezing. Remove chilled carcasses from ice and water, hang by the wing, and let drain 10 to 30 minutes before bagging and placing in storage.

Refrigeration: Refrigerate at 29 to 34° F. or freeze. Fresh-dressed ready-to-cook poultry should not be kept over five days in a refrigerator. If poultry is to be frozen, this should be done by the third day after it is dressed and chilled.

Freezing: Freshly dressed poultry should not be frozen until after it is chilled to 40° F. or below. Do not put warm freshly dressed poultry in the freezer.

Chicken barbecue style

Summer time, party time, anytime is a good time for serving barbecued chicken. Chicken is fun to cook and fun to serve for any group, so give

it a try the next time you serve. Knowing a few "tricks of the trade" will help make your barbecue even better.

The grill on which chicken is barbecued can be just about any style. It should be selected for versatility and to give adequate grill space for cooking, according to the size of group being served. For most backyard barbecues a charcoal grill that may already be on hand or available at a local store is satisfactory. If a larger grill is required, one constructed from small or medium sized metal drums split lengthwise works well.

Racks or grills to fit either the metal drum or a pit can be made by using a steel rod frame covered with 1″ x 2″ mesh welded wire. A sandwich-type rack allows all the halves to be turned at the same time. Pivot points on each side of the sandwich racks allow easy turning by one person. Not only is this type of rack a time-saver, but it also insures that all the chicken will be turned. The size of the rack depends, of course, on the size of the groups for which it will be used.

A rule of thumb is to allow one square foot for each four chicken halves. A maximum of three feet square could be used, but anything larger is too large to handle.

Fuel

Charcoal briquettes are the most convenient fuel for chicken barbecuing. They produce even heat without smoke. A good grade of hardwood charcoal is recommended for ease of starting and a long-lasting hot fire. Each half chicken being cooked requires one-half to three-quarter pound of charcoal. Approximately one-half pound can be used initially and the remainder added later if needed.

The charcoal should be piled in the center of a small grill when the fire is started. For pits, leave the charcoal in the bags. Cut the bags open on top and place them in the center of the pit. A commercial lighter fluid or kerosene can be used to start the fire. With fluids that have a strong odor, such as kerosene, be sure to allow sufficient time, usually 20 to 30 minutes, for all the liquid to burn off before starting to cook the chicken.

Gasoline is highly flammable, very dangerous, and should *never* be used for starting the fire.

After the briquettes are burning evenly, spread them over the bottom of the grill or pit. Garden rakes work very nicely for this task in large pits. Even distribution is important for uniform heating at all locations. During the cooking time, the briquettes should never be more than one layer deep.

Ready-to-cook broiler chickens weighing two to two and one-half pounds are ideal for barbecuing. Uniformity among all birds is desirable to assure that all pieces are ready to eat at the same time. One-half chicken should be prepared for each person being served.

Cooking the chicken

For large barbecues it is a good idea to load the turning racks while the fire is starting. By having all the chicken racked before the fire is ready, you can place all of it over the coals at the same time, and then all should be done at the same time. Starting all chicken in the same position also makes turning it easier. Chickens should be placed on the grill bone side down. If most of the cooking is done on the bone side there will be fewer problems with burning.

Pieces should be placed close together to conserve heat and grill space. For best use of grill space, arrange the halves in alternate rows of left and right halves. Be sure not to overlap, or cooking will be uneven.

For best results on home type grills, keep chicken as far from the heat as the construction of the grill will allow. A distance of six to eight inches works best, and the heat must be carefully regulated to prevent burning. With the larger grills or pits about 20 to 24 inches is the recommended distance between the charcoal and the chicken. This distance prevents the chicken from being cooked too rapidly or becoming charred and burned on the outside. Slow cooking gives best results, so during the cooking process regulate the heat carefully. If there is too much heat, sprinkle water on the charcoal, or if more heat is needed add more charcoal.

Turn chicken about every five minutes throughout the cooking period depending on the speed of cooking. Use tongs in turning small numbers of pieces where turning racks are not being used. Forks should never be used for turning because they pierce the skin and allow moisture to escape. Use clean white gloves if you wish to turn chicken by hand.

If the barbecue sauce does not contain enough salt, sprinkle salt on the chicken during the cooking process. Salting a small amount several times during the cooking usually produces a tastier product. Salt both sides of the chicken half, using approximately one-half teaspoon salt per half.

Test for doneness

Allow plenty of time for the chicken to cook (one to one and one-half hours depending on the size of chicken). Raw chicken can spoil the party. An easy test for doneness is twisting the leg bone or examining the wing joint. If the leg or wing bone twists free at the joint with moderate pressure, the chicken is done.

Sauce

Barbecue sauce should be added sometime after the birds have begun to cook, at least no later than 30 minutes before the chicken is done. The purpose of the sauce is twofold: It adds flavor, and it prevents the chicken from becoming dry and tough. Sauce may be applied with a pastry brush, a small "mop" on a stick, a new dish mop, or a new paint brush.

Many different kinds of barbecue sauces are available and, if used properly, will give satisfactory results. However, one disadvantage and a good reason not to use those containing tomato products is that burning or charring easily occur. If sauces containing tomato products are used, they should be used only at the end of the cooking period. An oil and vinegar sauce can be used during most of the cooking time, then the tomato sauce later.

Chicken should be served while hot. Holding chicken in foil-lined boxes or pans will keep it hot if serving is delayed.

Recipe for barbecue sauce

Ingredients

Number of Halves	10	50	100	250
Butter or margarine	⅛ lb.	¾ lb.	1 ¼ lb.	3 lbs.
Water	1 cup	5 cups	10 cups	7 qts.
Vinegar	½ cup	2 ½ cups	5 cups	3 ½ qts.
Dry mustard	¾ tsp.	1 ½ tbs.	2 ½ tbs.	6 ¼ tbs.
Sugar	1 tbs.	5 tbs.	10 tbs.	1 ⅔ cups
Salt	½ tsp.	5 tsp.	10 tsp.	½ cup
Chili powder	½ tsp.	2 ½ tsp.	5 tsp.	¼ cup
Black pepper	½ tsp.	2 ½ tsp.	5 tsp.	¼ cup
Paprika	½ tsp.	2 ½ tsp.	5 tsp.	¼ cup
Onion powder	½ tsp.	2 ½ tsp.	5 tsp.	¼ cup
Garlic powder	⅛ tsp.	⅝ tsp.	1 ¼ tsp.	1 tbs.
Worcestershire sauce	1 ½ tsp.	2 ½ tbs.	5 tbs.	6 oz.
Tabasco sauce	1 ½ tsp.	2 ½ tbs.	5 tbs.	6 oz.

Combine the butter, water, vinegar, Worcestershire sauce, and Tabasco sauce and bring to a boil. Combine the dry ingredients and add to boiling liquid. Boil gently for one hour. The amount given is for medium-sized birds. Plan for more sauce if larger birds are used. For small family barbecues the sauce may be brushed on with a pastry brush or similar applicator. For larger numbers the sauce is more easily applied by being sprayed on the birds, using a stainless-steel sprayer, usually three- or three-and-one-half-gallon capacity. The sauce should be kept hot and stirred frequently to keep an even mixture.

Chapter 11

Housing for your home poultry flocks

WHEN I WAS A SMALL boy, my folks got serious about the chicken business. They set out to do it the right way on what then was considered a commercial scale.

The laying house eventually was built. It had a sound concrete foundation, adjustable windows for ventilation, screening, and electricity for the 14-hour working day. No one thought about running water or a feed storage room inside the building. It was built for hand carrying. We used corn cobs for bedding. The young pullets with their bright combs made a pretty sight as they took up residence in their new home.

Your family poultry flock requires a sound shelter but it doesn't need to be fancy. Your first goal should be to make use of existing buildings if possible. Your flock is not going to be a big money-maker. That means that you probably can't justify building brooder and laying houses on strictly economic lines. But if your family has decided that you want the advantages of fresh eggs and poultry meat, then you will need to invest in the necessary structures.

If you buy started pullets or yearling hens, you will need only one structure. One is sufficient for a broiler flock, too. But if you plan to brood and rear your own pullets you will need two houses. One will be the brooder and the other the house for the mature hens that are providing your family with eggs. It is a good idea to have the two structures separated from each other so that the new flock does not pick up disease and other pest problems from the older birds.

Housing for commercial poultry flocks is so well-designed with insulation and mechanical ventilation that birds are comfortable except for a few very warm days each summer. Heat from the birds warms the house even with some ventilation in cold weather and keeps room temperatures about 55° most of the time and always safely above freezing.

Whenever possible the beginning hobby flock should use existing housing with little or no additional cost. That must not, of course, extend to the point of cruelty to animals through unreasonable exposure. Birds must be protected from storms and from temperature extremes both from a humane standard and to permit them to remain healthy and productive.

Too much space can be a problem in cold weather. Ten or 15 birds need only 30 to 40 square feet of floor space, less if kept in multiple-deck cages. It may take some ingenuity to design an area within an existing shelter that will allow birds to keep themselves warm safely. Use of windows for light is not essential. A 15- or 25-watt bulb will suffice, a timer being used to provide 14 hours of light per day for layers.

Build a new structure?

The cost of a new building for the hobby flock may be prohibitive. Where cost is not a strong factor, copies of small shed-type houses can be built by flock owners with average skills. Plans are on the following pages. Use of exterior-grade plywood makes the building strong and easier to build. The plywood can be used for floor, exterior siding, inner lining, and roof deck. It must be of the proper grade for the intended use. The building can be attached to skids so that it can be moved around, or it can be built permanently with a concrete floor and foundation.

Wooden-floored buildings should be raised on blocks, with the floor about 12 inches off the ground to avoid dampness and to discourage rodents. Pole buildings with dirt floors are sometimes used, but they have difficult rodent problems and are hard to clean properly.

Compact hutch-type structures designed to be serviced from the outside are sometimes used for the small flock. For seasonal use only, they can be a wire-floored, open design with nests, feed, and water outside the pen area. It should be possible to plan a convertible arrangement that would allow for fitting insulated panels on the sides and the floor so that the birds would be comfortable all year.

Ventilation is important

Provisions for ventilation and insulation to temper the extremes of heat and cold should be considered. A single-walled uninsulated structure will be cold and damp in winter and too hot in summer. Ceiling insulation is most important. Sidewall insulation is desirable. Ventilation by windows or slot vents that can be adjusted to the weather is necessary in a tightly constructed house.

Housing should be adequate to provide comfortable conditions so that the birds will stay healthy and productive whether confined year-round or permitted some range during mild weather.

The basic requirements of a poultry house are that it provide sufficient floor space, protection from weather and predators, and be well-ventilated but free from drafts.

With floor-type operations the layer house should provide at least two square feet per bird for light breeds such as Leghorns and three square feet per bird for the heavier breeds.

Cover openings such as windows with one-inch mesh poultry netting. During cold weather the openings can be partially or completely covered with polyethylene film if needed, but be sure to provide adequate ventilation.

Layer cages are excellent for a small home flock. Cages can be installed in any small building, or a shed-type building can be constructed especially for cages.

Plenty of equipment

The layer house must be furnished with the following equipment: feeders, waterers, nests, and lights. Besides these items, the furnishings might include containers for granite grit and oystershell, a droppings board or pit, and a roost.

Clean, fresh water should be available at all times. If trough-type waterers are used, provide at least one linear inch of waterer space per layer. If fountain-type waterers are used, provide enough fountains to have a capacity of two to three gallons of water for each 25 layers. This will assure you of adequate water supply for 24 hours. Adjust height of waterers in the same manner as feeders. Fountain-type waterers can be placed on wire platforms about four inches high to help prevent water spoilage and contamination of the water.

You have a choice on selection of equipment. Time-tested units are available at moderate prices from mail order catalogs and from feed supply stores. You can build feeders, roosts, and other equipment with wood-working tools. Waterers probably should be purchased.

Good nests save eggs

An adequate number of properly constructed nests will result in cleaner eggs and fewer broken eggs. Nests can be arranged in the center of the house or along the walls. Hens prefer a darkened area for nesting.

Nests may be individual nests designed to accommodate one hen at a time or community nests designed to handle several hens at once. Individual nests are usually 10 to 14 inches wide, 12 to 14 inches high, and 12 inches deep. Individual nests are usually arranged in rows several tiers high. Have perches below the entrance of the nest with the lowest nest being 18 to 20 inches above the floor. Provide an individual nest for every four layers.

Community nests are used by several layers at one time. These nests are usually about four feet wide and two feet deep and constructed similar to the one shown on page 122. One community nest this size is adequate for a family flock up to 35 hens.

Droppings boards or pits and roosts are not considered to be necessary, but they do help prevent manure buildup in the litter. Roosts are usually placed over the droppings board or pit. Roost poles can be made from two-inch by two-inch pieces with the upper edges slightly rounded. Roost poles should be placed 14 to 15 inches apart and should provide 8 inches of linear space per bird. Roosts should be 16 to 24 inches above the floor with one-inch by two-inch welded wire below the roost to keep hens out of the droppings.

Cage houses

Equipment for cage houses includes feeders, waterers, and lights. Commercial feeders and waterers are normally provided with the cages when they are purchased. If the cages are in an open-type house, use curtains or polyethylene film on the sides of the house during the winter months. However, allowance must be made to provide adequate ventilation.

Litter in layer house

Start with four inches of a moisture-absorbent litter, such as wood shavings. Keep the litter dry and in a loose, uncaked condition. Wet litter causes dirty eggs and increases the risk of disease problems. Preventing water spillage and leaks and providing proper ventilation will help keep the litter in good condition. Stir the litter when it becomes damp and packed. Remove any wet spots and replace with fresh litter.

A basic poultry structure

This eight- by eight-foot poultry house is a basic design from University of Wisconsin that can be used for either a brooder house or a laying house. It also can be adapted for bantams, pigeons, or rabbits if the family should decide to drop the poultry flock. The house requires basic lumber and can be built by people with average handyman skills.

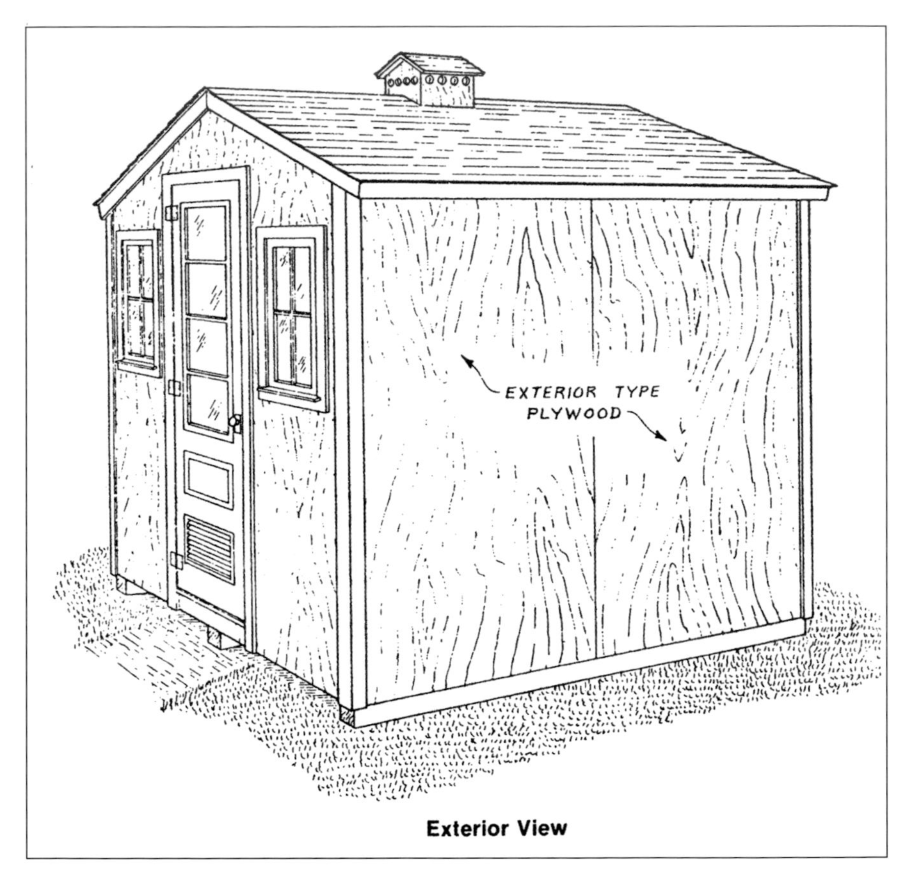

Design: Agricultural Engineering Department, University of Wisconsin, Madison, 53706

Interior for brooding chicks

This plan shows the basic house arranged to brood chicks. Heat lamps hung from the ceiling provide warmth. Bendable material keeps the chicks out of corners where crowding could be a problem.

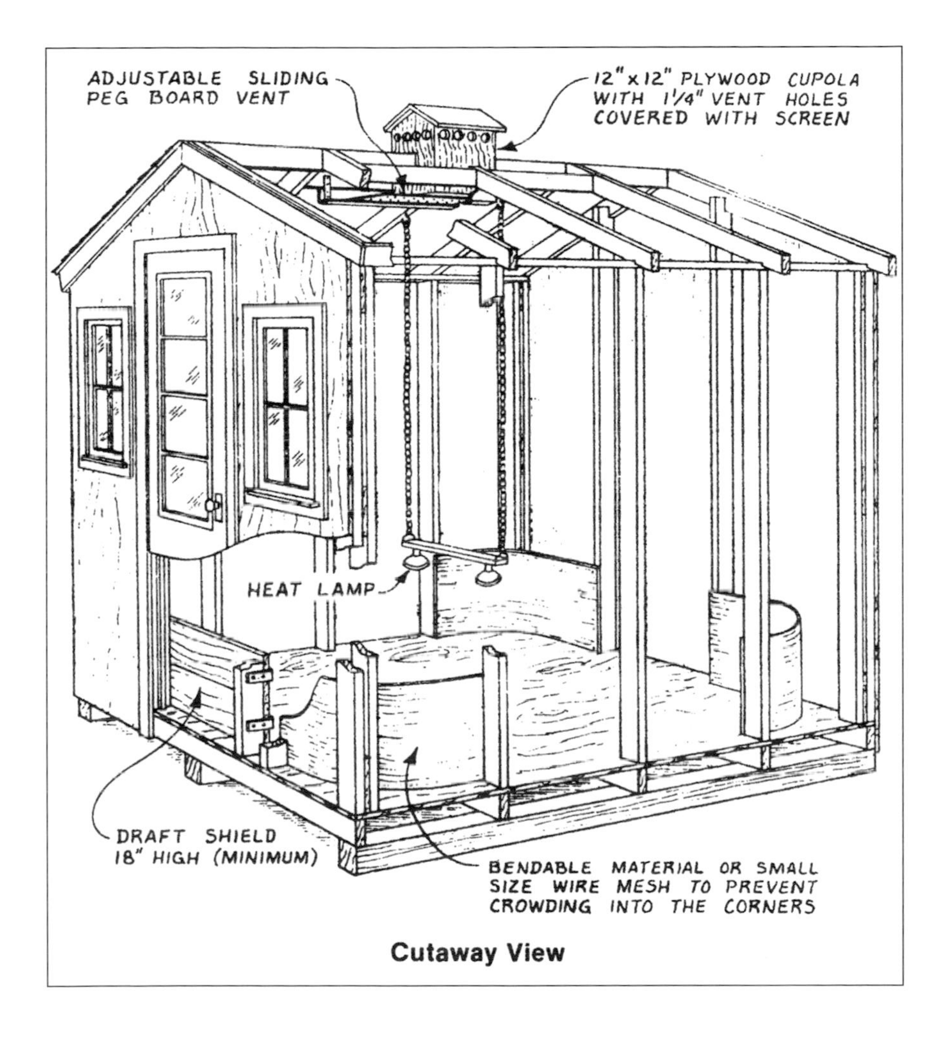

Cutaway View

Interior for the laying house

The basic house will house 15 to 20 hens comfortably. The nests and roost bar are installed along the back wall. The cupola with sliding vent helps control the temperature.

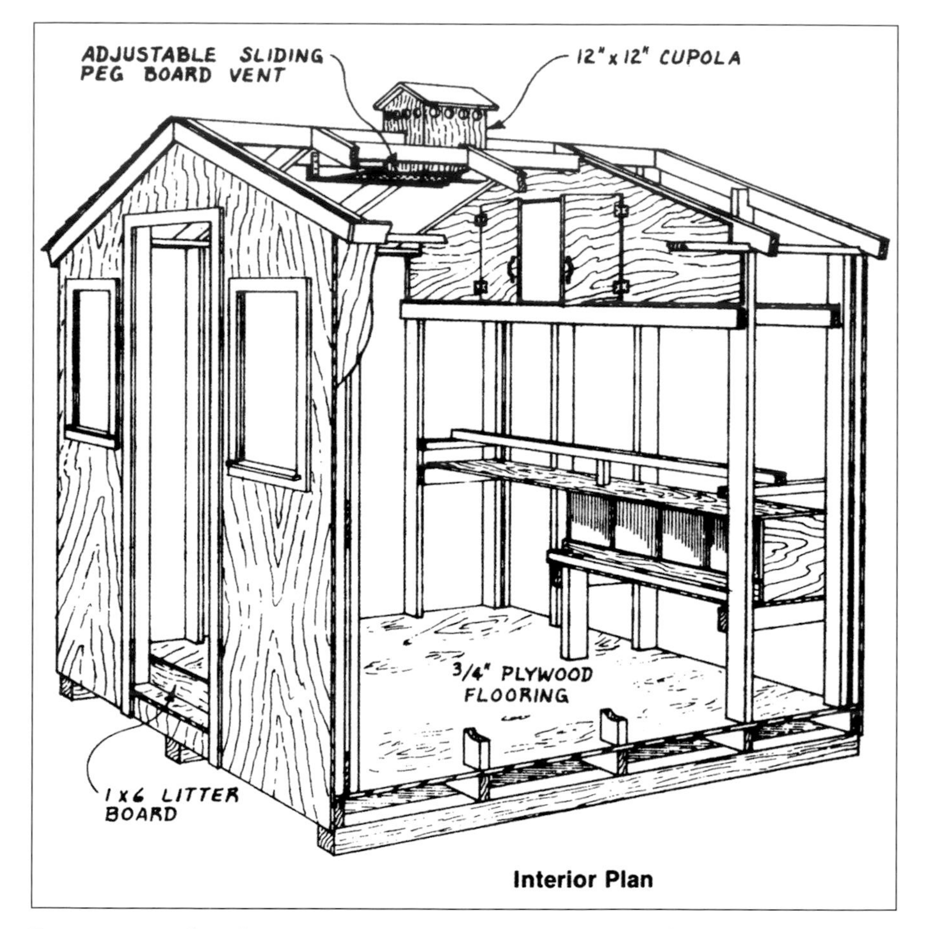

Interior Plan

Design: Agricultural Engineering Department, University of Wisconsin, Madison 53706.

Side-section view of laying house

This view shows a side cut-away of the laying house. Note how the studs and roof are arranged. The basic structure is the same for the brooder house version. Walls are made with four-by-eight sheets of exterior plywood.

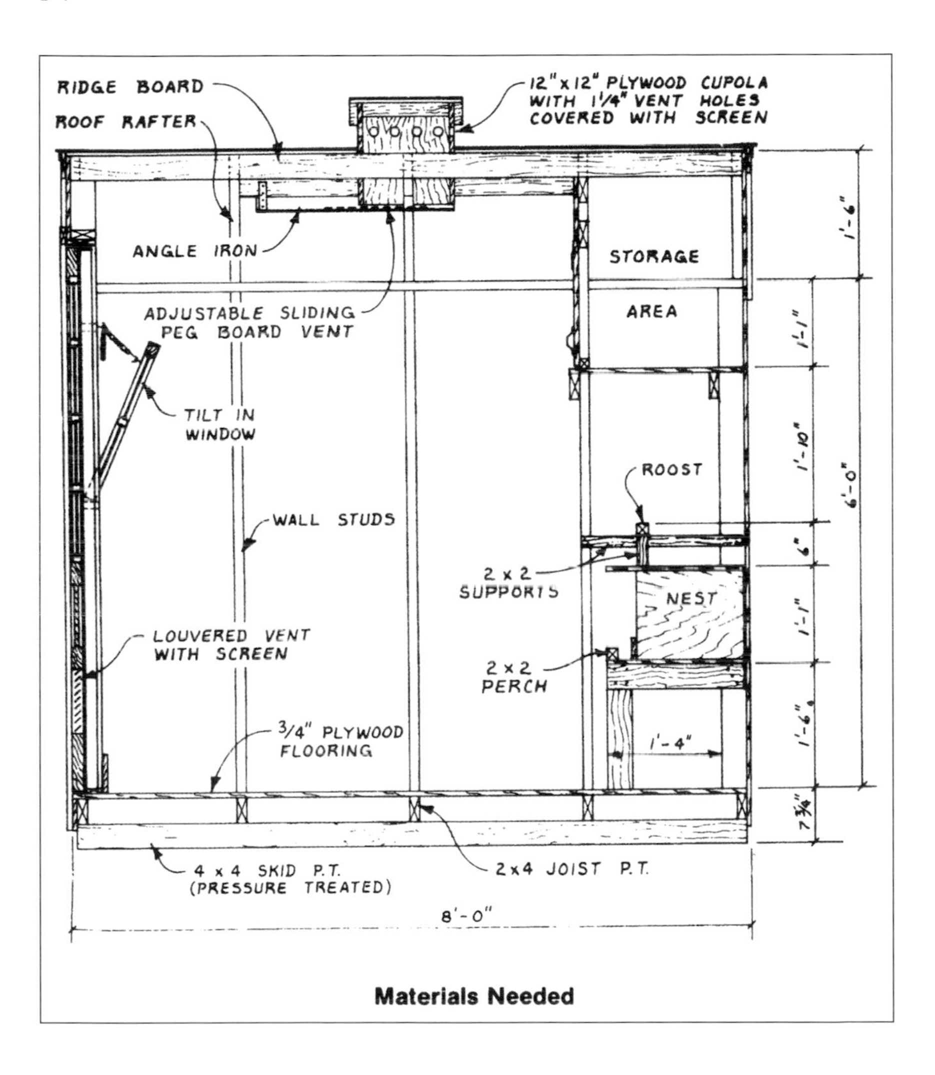

Materials Needed

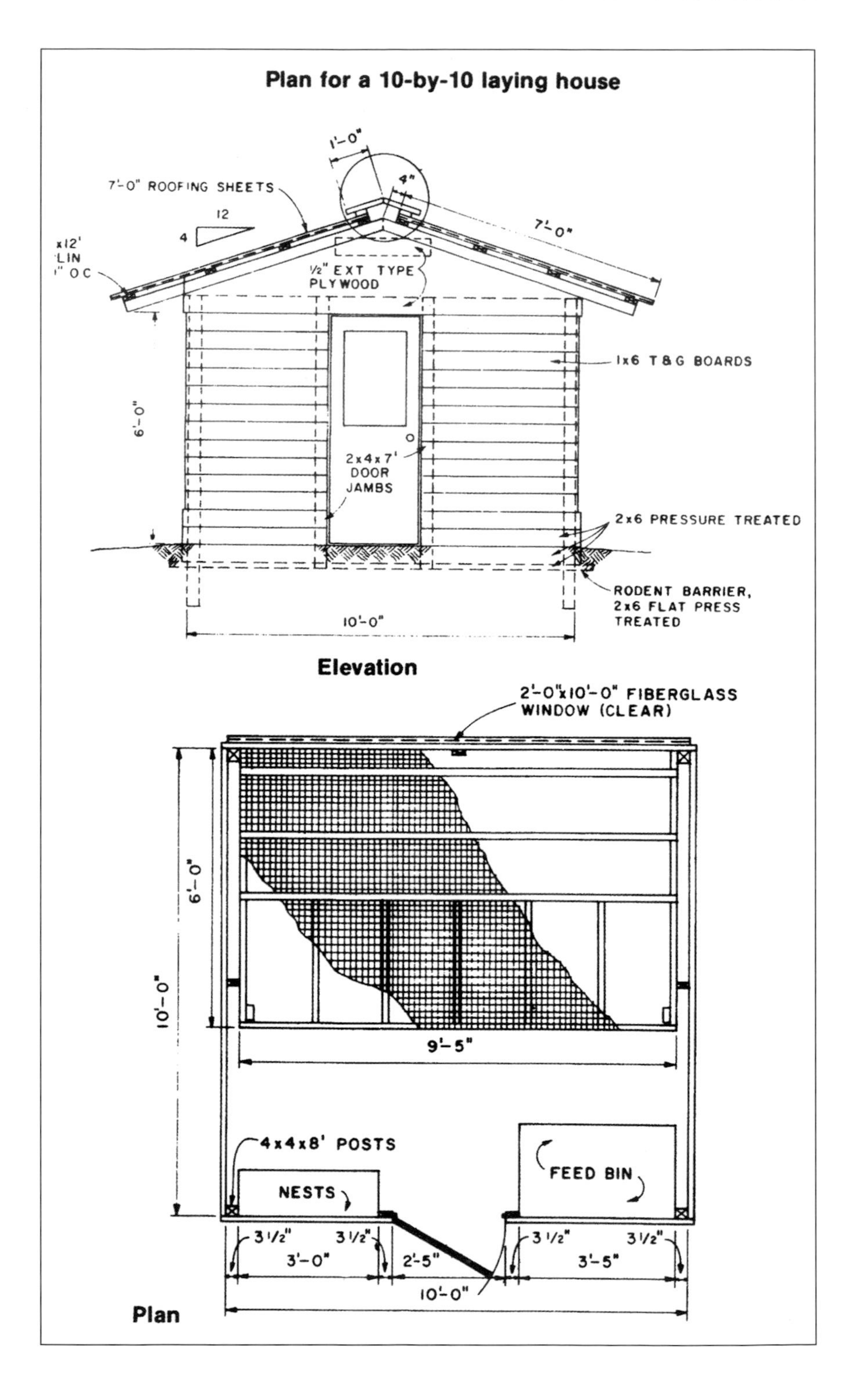
Plan for a 10-by-10 laying house
1'-0"
4"
7'-0" ROOFING SHEETS
12
4
7'-0"
x12'
LIN
1" OC
1/2" EXT TYPE
PLYWOOD
1x6 T&G BOARDS
6'-0"
2x4x7'
DOOR
JAMBS
2x6 PRESSURE TREATED
RODENT BARRIER,
2x6 FLAT PRESS
TREATED
10'-0"
Elevation
2'-0"x10'-0" FIBERGLASS
WINDOW (CLEAR)
6'-0"
10'-0"
9'-5"
4x4x8' POSTS
NESTS
FEED BIN
3 1/2"
3 1/2"
3'-0"
2'-5"
3 1/2"
3 1/2"
3'-5"
10'-0"
Plan

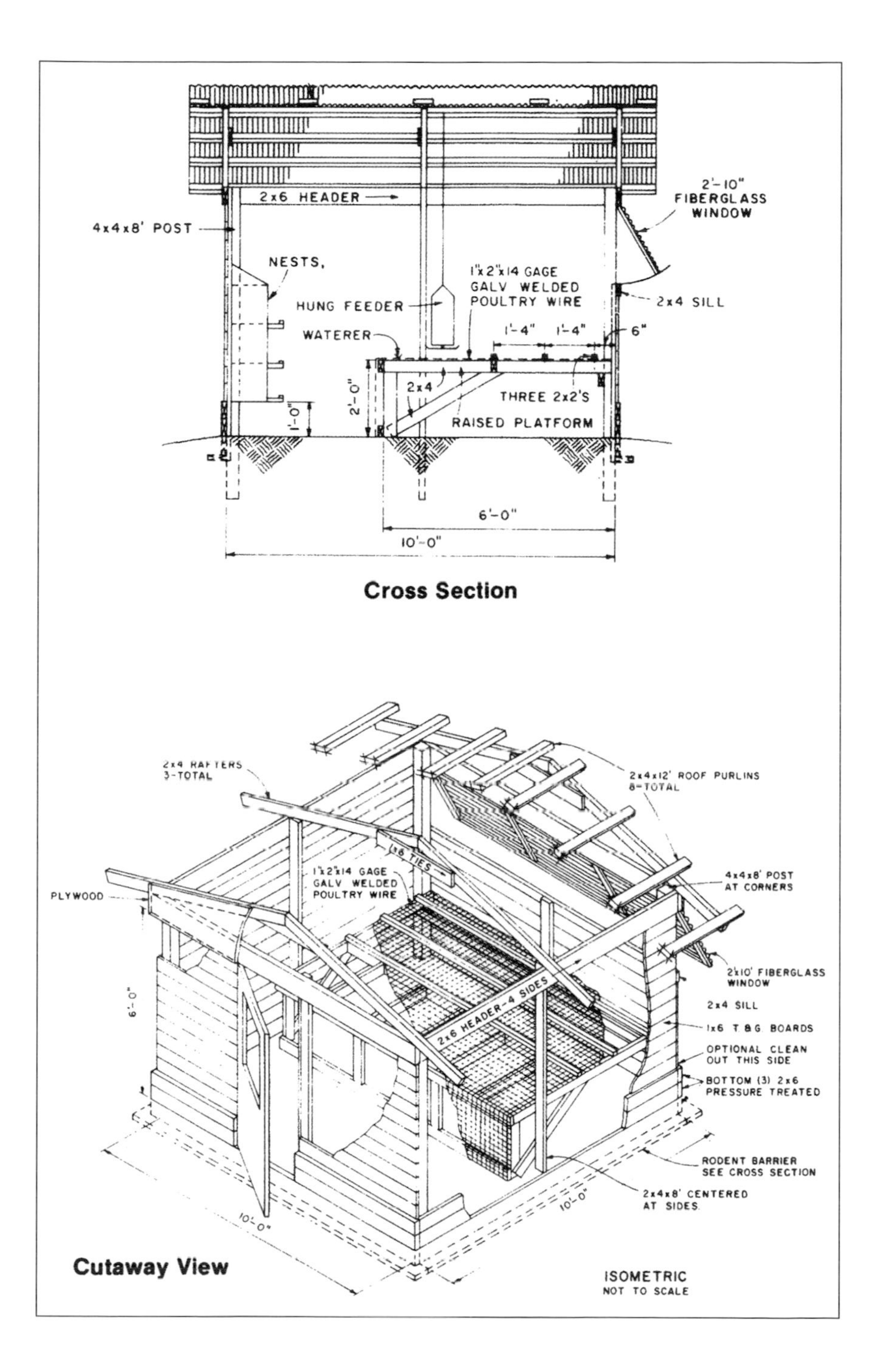

Cross Section

Cutaway View

A brooder with a sun porch

This brooder is simple to construct and relatively inexpensive. It can rear 50 chicks up to broiler size. There's enough brooding capacity for 150 chicks. The porch can be enlarged if you want to rear more than 50 meat birds. Four 40-watt electric bulbs provide brooding heat. Cover wire mesh floor with papers during the first week to ten days of brooding.

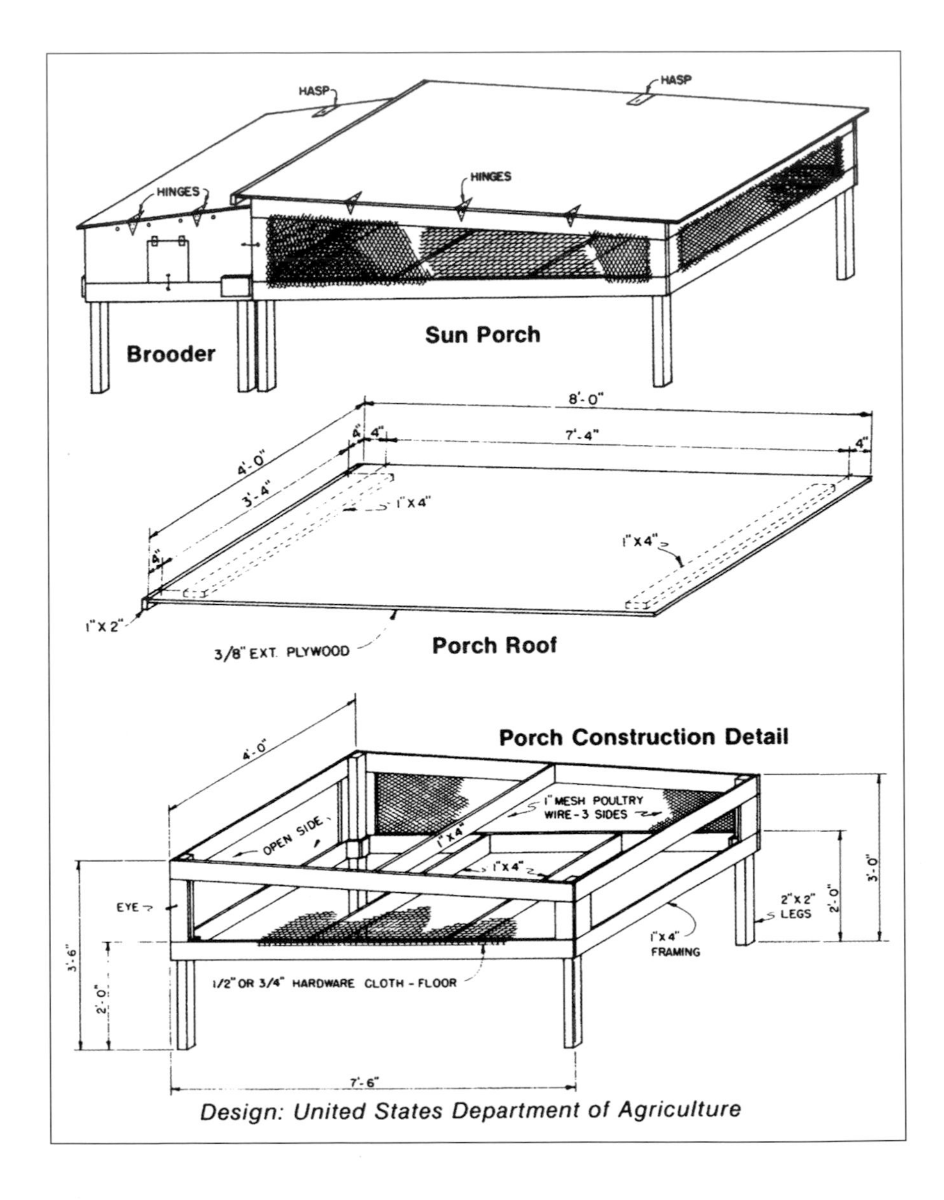

Design: United States Department of Agriculture

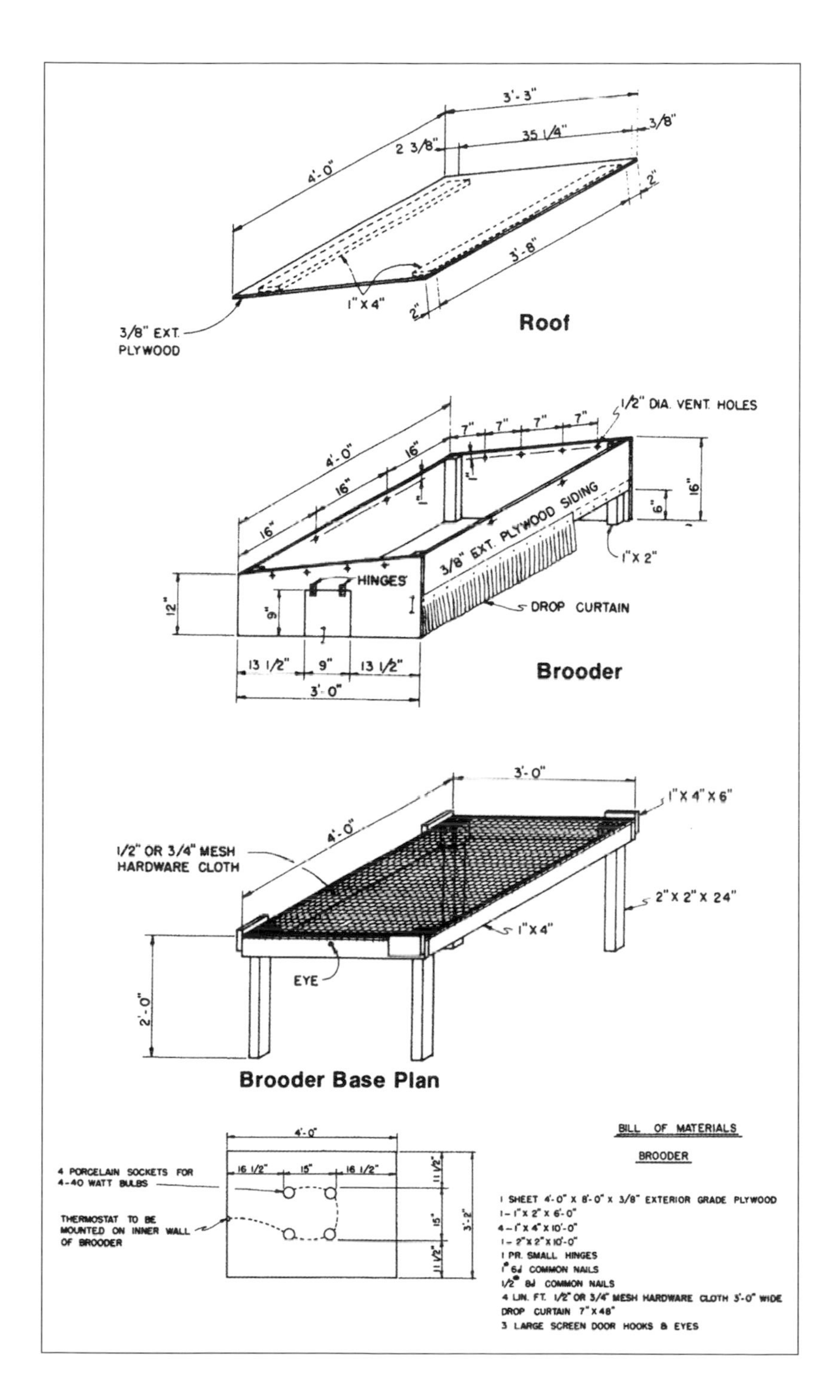
Roof
3/8" EXT. PLYWOOD
1" X 4"
Brooder
1/2" DIA. VENT. HOLES
3/8" EXT. PLYWOOD SIDING
HINGES
DROP CURTAIN
1" X 2"
Brooder Base Plan
1/2" OR 3/4" MESH HARDWARE CLOTH
1" X 4" X 6"
2" X 2" X 24"
1" X 4"
EYE
4 PORCELAIN SOCKETS FOR 4-40 WATT BULBS
THERMOSTAT TO BE MOUNTED ON INNER WALL OF BROODER
BILL OF MATERIALS
BROODER
1 SHEET 4'-0" X 8'-0" X 3/8" EXTERIOR GRADE PLYWOOD
1 - 1" X 2" X 6'-0"
4 - 1" X 4" X 10'-0"
1 - 2" X 2" X 10'-0"
1 PR. SMALL HINGES
1" 6d COMMON NAILS
1/2" 8d COMMON NAILS
4 LIN. FT. 1/2" OR 3/4" MESH HARDWARE CLOTH 3'-0" WIDE
DROP CURTAIN 7" X 48"
3 LARGE SCREEN DOOR HOOKS & EYES

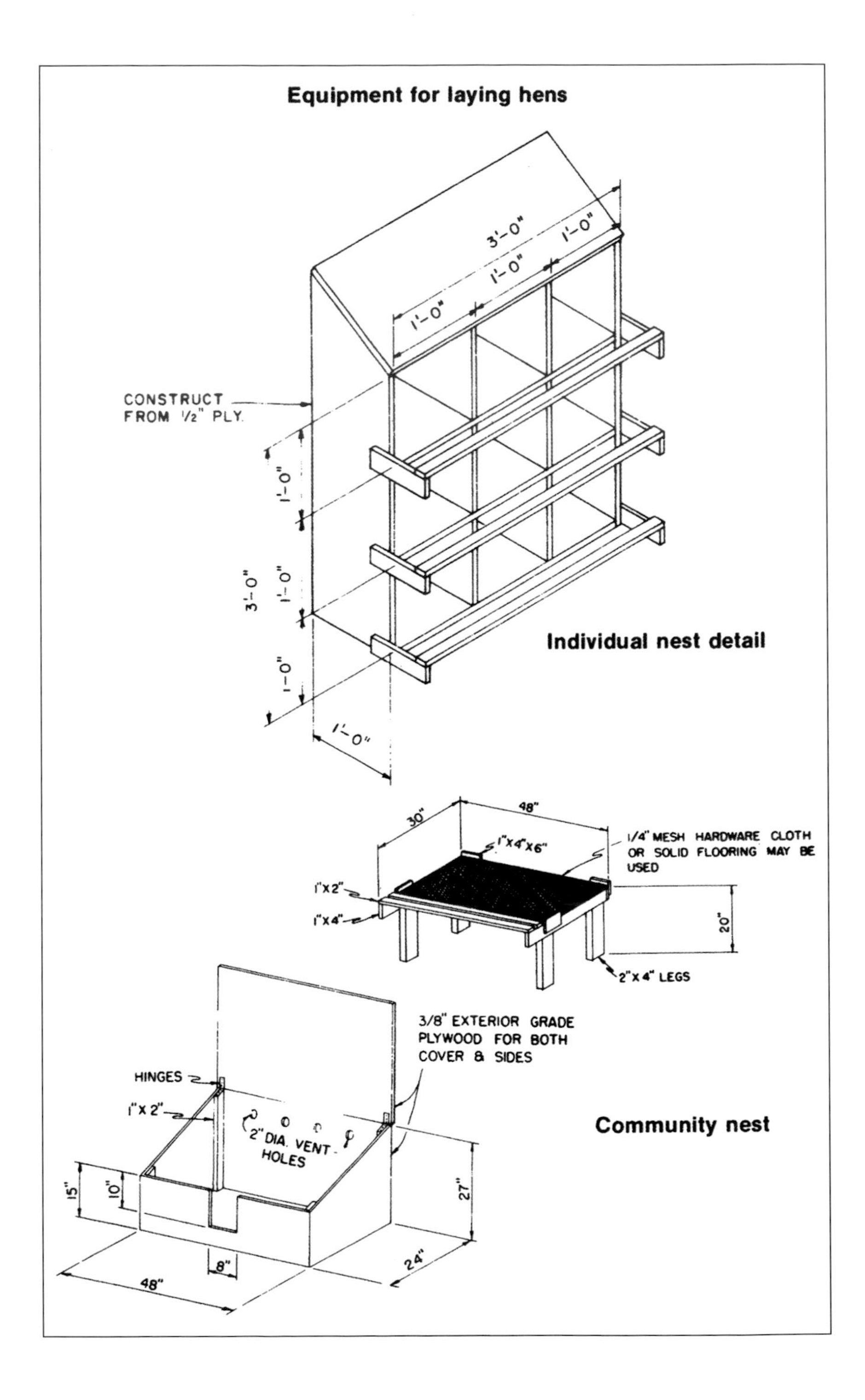
Equipment for laying hens
3'-0"
1'-0"
1'-0"
1'-0"
1'-0"
CONSTRUCT FROM 1/2" PLY.
1'-0"
3'-0"
1'-0"
1'-0"
1'-0"
Individual nest detail
48"
30"
1"x4"x6"
1/4" MESH HARDWARE CLOTH OR SOLID FLOORING MAY BE USED
1"x2"
1"x4"
20"
2"x4" LEGS
3/8" EXTERIOR GRADE PLYWOOD FOR BOTH COVER & SIDES
HINGES
1"x2"
2" DIA. VENT-HOLES
Community nest
15"
10"
27"
8"
24"
48"

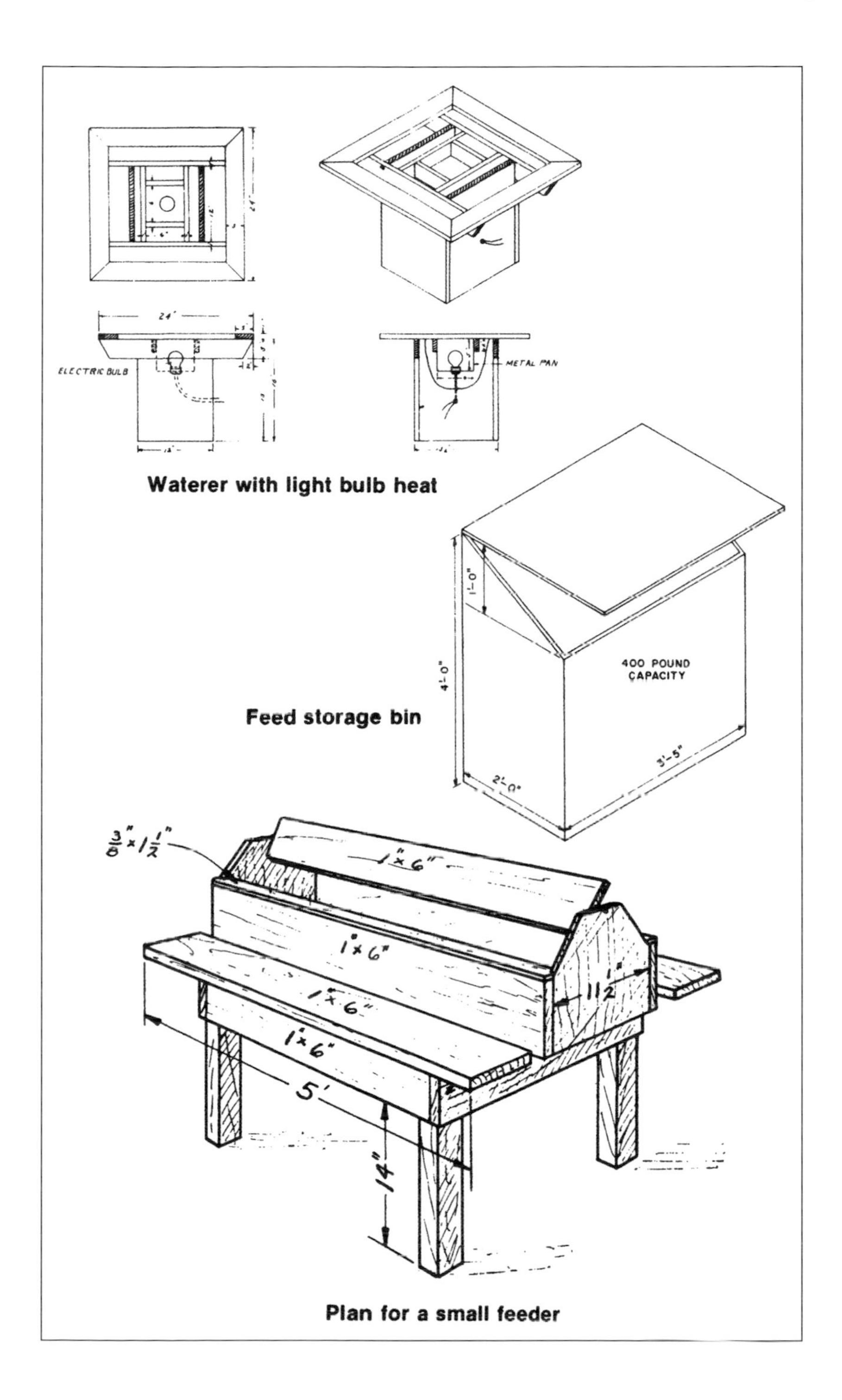

Waterer with light bulb heat

Feed storage bin

Plan for a small feeder

Chapter 12

How to fight disease and pest problems

WE HAD OUR SHARE OF disease and pest problems around our chicken houses. In those days, a sick chicken either got better or it didn't. Today, there's a drug for almost every ailment . . . if you can figure out the problem and take action immediately. The basics haven't changed. Good feed, plenty of water, and cleanliness are the keys to healthy birds, just as they were a generation ago.

The episode of the skunk is the pest problem I remember best. This skunk got into the brooder house and had a chicken dinner. Dad took a quick look and went for the shotgun. He blasted a hole about three inches wide through the back wall, right on target. The skunk departed this world in spirit but the reminders of its body stayed on for weeks.

Another incident was the case of the human predators. Chicken thieves were active because a bird could be turned into quick cash in

50 different places around the county. One summer night we heard the squawk of pullets being pulled from their perches in the trees. Once again Dad got out his trusty shotgun but this time the "skunk" got away.

Sooner or later you will encounter disease and pest control problems with your family flock. Chickens have been selected to resist disease through many generations. However, disease does strike from time to time. External parasites, mice, and rats can be problems, too.

A well-fed flock, housed in comfortable surroundings, will have a minimum of these problems. Good management on a day-to-day basis is the key to poultry health and well-being.

Large-scale poultrymen have come to anticipate some losses. During the first three weeks of a chick's life, they expect around 2 percent of the birds to die. If you start 50 chicks, normal mortality would be one. After the first three weeks, mortality should not be greater than 1 percent per month. There may be a surge of problems as adult flocks come into egg production.

Recognize the disease early. Look for unhealthy birds. This may help keep the disease from spreading and may reduce cost of treatment.

Check your flock every day. Note the birds' actions and how they are eating and drinking. Listen for any unusual sounds—any sneezing or rattling.

Recognizing a disease in its early stages is aided by record keeping. A slump in feed and water consumption is usually one of the best early indicators. It pays to keep daily records on feed and water consumption, egg production, and mortality. Any major change from day to day, or over a period of time, may mean that a disease is present in the flock.

All chickens are susceptible to disease. Infections may occur in single individuals or be widespread depending upon the infectious agent involved, the resistance of the flock, and the environment provided. Disease prevention depends on four major considerations:

1. Cleanliness
2. Proper nutrition
3. Proper environmental conditions
4. Presence of some immunity (vaccination programs)

Cleanliness is a must

There is no substitute for cleanliness. This includes everything the birds or eggs come in contact with either directly or indirectly.

Clean, disinfect, and air out a poultry house before putting birds of any age in it. Disinfecting is not a substitute for cleaning. Disinfectants are only effective on clean surfaces. To prepare a previously occupied poultry house, follow these steps:

1. Remove dust, dirt, crusted manure, litter, feed, etc.
2. Thoroughly clean the building and all equipment, including air intakes, overhead ledges, and fans. High-pressure water (600 to 1,000 pounds per square inch) is best. Chemicals added to the water aid cleaning and disinfecting, but high-pressure water alone is excellent.
3. Follow the manufacturer's instructions if disinfectants are used. All disinfected surfaces must be dry and the building aired out before birds are put in.

Clean automatic waters daily: Slime and decomposing feed can harbor disease organisms and molds.

Nutrition: Nutrition plays a major role in the health of the flock and in its ability to resist disease. The nutritional needs of a healthy flock are normally satisfied by an adequate supply of clean, fresh feed, easily accessible, and containing the necessary amounts of required ingredients.

Tonics, appetizers, and stimulants are not necessary. Use them only under the direction of a qualified poultry specialist.

Environment: Environmental conditions contribute directly to flock health. Poor environment makes birds more susceptible to disease as well as contributes to the spread of many diseases.

Adequately lighted, well-ventilated, relatively dry quarters are desirable. A good environment provides:

1. Shelter, protection, and comfort.
2. Convenient and adequate supplies of proper feed, clean water, and fresh air.

3. Equipment and facilities that are conveniently arranged for both birds and caretaker.

Waterers should be easily reached by the birds, of adequate capacity, and should *not overflow.* Feeders that waste feed or become fouled aid the spread of disease.

If litter is used, it should be absorbent, relatively dust-free, and resistant to matting. Litter under waterers and feed troughs should be replaced when it becomes damp or matted. Moisture in litter fosters parasite development and growth of bacteria.

With cage systems, the birds are unable to move to a more acceptable or healthy location. The manager must control the environment uniformly to get satisfactory performance.

Vaccination programs: Immunization by vaccination is recommended for some diseases. Planning is important because immunity is not effective immediately. When planning vaccination programs consider the disease history of your place, the community, and the state or region—and vaccinate only healthy birds.

The following protection program is approved in many states:

1. Vaccination is generally recommended for bronchitis and Newcastle.
2. Vaccination is conditionally recommended (depending on history of premises and community) for fowlpox, laryngotracheitis, epidemic tremors, and cholera.
3. Inhibitory drugs in feed are generally recommended for coccidiosis if brooding is done using the floor system. If the cage system is used for brooding, inhibitory drugs may not be needed.

When administering vaccines or drugs, follow the manufacturers' directions precisely.

Key points in disease prevention

1. Keep visitors away from poultry buildings.
2. Do not mix birds of different ages.
3. Remove sick, injured, and dead birds as soon as noticed, and burn or bury dead birds immediately.

4. Screen or cover all poultry house openings so wild birds and rodents cannot enter.
5. Clean trough-type waterers daily. Check operation of individual cups and direct action valves or nipples. At the same time, check operation of feeders, fans, and lights.
6. Remove wet litter spots as they occur.
7. Never permit contaminated equipment (crates, tools, trucks, etc.) from other poultry farms in your buildings.
8. Obtain reliable diagnosis before administering drugs or biologics.

Intestinal parasite control

Piperazine is the recommended treatment for round worms, i.e., the large intestinal round worms.

Hygromycin offers the most satisfactory control of capillaria worms.

A specific control for tapeworms is not known. Some companies have preparations that may help, but do not always give the expected results. Composition of the various wormers differs, so follow the manufacturer's directions.

Worms are not usually a problem for flocks grown and maintained on wire or slat floors. However, all flocks should be periodically checked for the presence of worms.

External parasite control

Several kinds of lice and mites attack chickens. Heavy infestations cause discomfort and general unthriftyness. Lice found on birds, called biting lice, feed only on skin scales and feathers. They do not pierce the skin to suck blood. However, their irritation may lead birds to scratch, causing skin abrasions and open wounds.

Mites, microscopic relatives of the spider, pierce the skin and feed on blood. The Northern Fowl and Red Mite are two of Wisconsin's most common kinds. Red Mites generally attack birds at rest and live in cracks in the roost, floor walls, or other sheltered area. Northern Fowl Mites spend their entire life on the bird, congregating around the vent, at the base of new feathers, and anywhere the skin is rough. A heavy population can reduce the male's willingness to mate. Mites or blood spots may appear on the eggs.

Lice and mite control is simple if you use the right insecticide properly. Community-penned birds can treat themselves with an approved insecticide mixed with dusting material. If this is not practical, spray or dust each bird individually. For Red Mites, you only have to treat roosts, walls, litter, nest boxes, and other equipment. Treat all mite hiding places thoroughly. Insecticides usually do not kill mite and lice eggs. Apply a second treatment ten days after the first to kill newly hatched mites and those that escaped the first treatment.

State and Federal agencies approve insecticides before they can be sold. However, regulations change each year. Check with your local extension office before buying and using an insecticide.

Fly control in poultry houses

Fly control is an integral part of poultry management. With today's concern about environmental conditions, fly control takes on added importance. Besides their ability to mechanically carry disease organisms, flies may be considered environmental pollutants just by their presence.

Several species of flies are common in and around laying houses. The most common are the house fly and little house fly.

The house fly breeds in moist, decaying plant material including refuse, spoiled grains and feeds, and in all kinds of manure. Although it breeds in poultry manure, this is not a favorite medium. For this reason, it is more likely to be a problem around poultry houses where general sanitation is poor. The house fly prefers sunlight and is a very active fly that crawls about over filth, people, and food products with equal disdain. It is, therefore, the most important species from the standpoint of spreading human and poultry diseases and fly-specking eggs. This fly is also the intermediate host for the common tapeworm in chickens.

The little house fly is generally somewhat smaller than the house fly, but the size difference is not enough to be a good distinguishing characteristic. The little house fly prefers a less moist medium than the house fly in which to breed. Poultry manure is preferred over most other media. It prefers shade and cooler temperatures and is often seen circling aimlessly beneath hanging objects in the poultry house, egg room, and feed room. This fly is less likely to crawl about on people and food than is the house fly. On the other hand, it is usually the one that causes people living near poultry establishments to complain about a fly problem. Because of its preference for shade, it may collect in large numbers in nearby garages, breezeways, and homes. House flies and little house flies are capable of movement up to 20 miles from the site of development, but normally they move no more than a mile or so from this locality.

Blow flies (green or blue bottle flies) may occur in poultry houses. They breed in decaying animal carcasses including dead birds, dog manure, broken eggs, and wet garbage. Any reasonable sanitation program is usually sufficient to hold them in check.

Soldier flies are large (about twice the size of a house fly), blackish flies commonly found around poultry manure. They are not pests in that they use the manure only as breeding medium and do not bother anything else. They may even be considered beneficial since their larvae appear to render manure unsuitable for production of house fly larvae.

Fruit flies are sometimes produced in large numbers where there is a mixture of manure, wasted feed, and water. Nearby homes may become targets of these flies with resulting complaints to the poultryman. Elimination of large amounts of wasted feed and repair of leaking water systems usually solves this problem.

Biology of flies

All flies go through four life stages. These are the egg, larva, pupa, and adult fly. Eggs are deposited on the breeding media, and larvae or maggots develop in this moist or wet material until ready to pupate. Generally, the mature maggots crawl out of this material and seek a drier place to pupate. Here they form a brown seed-like puparium from which the adult fly emerges. This development from egg to adult fly can take place in as little as seven to ten days under ideal conditions.

Cultural-biological control

Management of manure in such a way that it is not attractive for fly breeding is the most effective means of fly control. Fresh poultry manure generally contains 60 to 80 percent moisture. Fly breeding in this material can be virtually eliminated by reducing its moisture content to 30 percent or less or by addition of moisture to liquify it. Drying usually is preferable since dry manure is more easily handled, occupies less space, and has less associated odor problems than does liquid manure.

Chemical control of flies

Insecticides should be considered as supplementary to sanitation and management measures aimed at preventing fly breeding. A number of insecticides are available under various trade names for use as: (1) residual sprays, (2) space sprays, mists, and fogs, (3) resin strips, (4) baits, and (5) larvicides. Often, the same insecticide can be used in more than one of these control methods. Talk to your feed dealer about recommended materials for poultry houses. Follow directions carefully. Approved materials will not result in egg or meat contamination or injure birds if directions for use are strictly followed.

Residual sprays are usually the most effective and economical for controlling potentially heavy populations of adult flies of any species present. Such sprays should be applied in the spring at the beginning of the fly season. As a rule, they will continue to kill flies up to eight weeks. It is important that such insecticides be applied to surfaces on which flies alight. Generally, the framework of the house, the ceiling, trusses, wires that support cages, electric light wires and fixtures, etc. are favorite gathering places for the house fly and the little house fly. The outside of the poultry house, particularly around openings plus any shrubs or weeds adjacent to the house should also be sprayed since blow flies tend to congregate here. Do not spray birds or contaminate feed or water.

Space sprays, mists, and fogs with quick knockdown but no residual action are advantageous where immediate reduction of an adult fly population is necessary. There are many machines on the market designed to produce the small spray particle size desired for this type of application. Space applications should be to the point of "filling" the room with mist or fog.

Resin strips containing 20 percent dichlorvos (DDVP, Vapona) are readily available. When used according to directions, they give off a vapor that kills flies in enclosed areas where there is little or no air circulation. Thus, they are most effective in places such as feed rooms and egg rooms adjacent to the main poultry house. They are not effective in poultry houses because of the greater air circulation. Use one strip for each 1,000 cubic feet of room space.

Baits may be used in either a liquid or dry form. They usually contain an insecticide plus sugar or some other attractant. Baits are most useful as a supplement to residual sprays. They alone cannot be expected to control fly populations. Commercial dry baits in granular form are readily available, but liquid baits will have to be prepared by the user. Liquid baits are usually more effective since they can be brushed upon a wide variety of fly resting surfaces, as well as being placed in flat containers usually used for dry baits. Liquid baits should be used as soon as mixed and not stored for later use. They have the disadvantage of collecting dirt and dust. All baits must be placed out of reach of the birds.

Larvicides are applied directly to the manure below cages, screen wire, or slats. This type of application is designed to kill fly larvae (maggots) developing in the manure. It is necessary that the insecticide penetrate the manure and contact the maggots to kill them. This is often difficult since the constant addition of fresh manure offers new breeding material free of insecticide. This type of application is best utilized when reserved for treatment of fly breeding spots not eliminated by normal cultural practices.

Control of rodents

Rats and mice kill young chickens, destroy eggs, eat or contaminate poultry feed, and damage buildings by gnawing wooden walls, foundations, and equipment. They also spread many diseases and parasites.

Losses from rats and mice in a family-sized flock may be substantial, including chickens, feed, and other destruction. The damage often goes unnoticed, because losses are gradual and because rats and mice seldom appear when a person is in the poultry house.

To make new buildings rat-proof, use concrete foundations and floors. Metal shields around doors and screens over small openings are effective in keeping rodents out.

In fighting rats and mice, get rid of their breeding and hiding places on the farm. Clean out trash, dumps, piles of old lumber or manure, and garbage. Find and block runs or burrows. Then use a poison.

Use of anticoagulant poisons

Warfarin, Pival, Fumarin, Diphacinone, and PMP are all known as "anticoagulant poisons" because they prevent the normal clotting of blood and cause rats and mice to die by internal bleeding. All of these poisons are available in the form of ready-made baits. Some may be purchased as concentrates for homemade mixtures.

A single dose of either of these anticoagulant poisons will not ordinarily kill unless a very large amount is consumed. They are most effective when taken in small amounts for more than five feedings one or two days apart. Thus, ample quantities of bait should be made available for two weeks or more, since some individual rats and mice may not eat it when first exposed. In most instances, however, a marked reduction in bait consumption and rodent activity occurs after about the third day of treatment. The control of mice may require a longer period of time than for rats.

Anticoagulant poisons are capable of killing cats, dogs, and other warm-blooded animals. Chickens show considerable tolerance to the effect of the poisons. The potential hazard is reduced by the requirement of multiple feedings. To lessen the chance of accidental poisoning, a bait material should be used that is acceptable to rats and mice, but relatively unattractive to humans and domestic animals. Secondary poisoning from eating anticoagulant-poisoned rats and mice is unlikely if cats and dogs are well-fed when bait is exposed. Baits should be placed under protective cover that will permit free access by rats and mice, but exclude chickens and larger animals.

Bait preparation

When mixed with food in the recommended proportions, these poisons are quite tasteless and odorless to rats and mice, nor do the rodents associate the effect of the poison with the cause. Thus, in preparing bait, directions should be followed and only the amount of poison added that is called for on the package.

Bait placement

For rat control, place bait at or near places where rats are accustomed to feed. Use shallow pans, preferably not more than one half-inch high. If the bait is in a bag, tack one end down and make a small cross slit in it. On the farm, two or three one-pound placements in a barn, one or two in a corn crib, and one in each of the other buildings will usually suffice. Since house mice have a very restricted range, place two- or three-ounce amounts at not more than twelve-foot intervals along walls. Remember that bait must be continuously available to rats and mice. Check placements every one or two days, replace bait as needed, dispose of unpalatable bait, and relocate stations if bait is not being taken.

Predators

Confinement rearing has greatly reduced poultry losses from hawks, owls, crows, foxes, skunks, weasels, and other predatory animals. These wild animals and birds often attack chickens on range. Young chickens are especially vulnerable.

To protect birds on range from predators, close shelters at night. Screen the space beneath the shelters. Install an electric wire slightly above ground level on the outside of the fence that encloses the range.

Provide for sanitary disposal

Why should poultrymen do a good job disposing of dead birds? It's an objectionable task, and many producers aren't equipped to do the job conveniently and effectively. In addition, proper disposal takes time and effort as well as costing money.

Yet, sanitary dead-bird disposal is a must. The reason is that it reduces the spread of disease. Many diseases are spread by healthy birds picking on dead ones. So for that reason alone, it's a good idea to get dead birds out of a flock.

But that's not all. Producers also have to worry about odors and nuisances. Neighbors are becoming less tolerant as rural areas become more urbanized. Flies and odors are no longer considered a part of "living in the country."

What is the best disposal system? No one system of disposal is best for all conditions. Incineration is the most sanitary. But a disposal pit or deep burial may also be suitable, and they cost less than incineration. Commercial pick-up works well for those producers who have this service available to them.

Incineration. A good incinerator is an excellent means of disposing of dead birds. It is preferred over a disposal pit or deep burial in areas with poor soil drainage or where there is a danger of contaminating the water supply. The main disadvantages are the initial cost of the incinerator plus the operating expense.

An incinerator should burn the carcasses to a white ash. Many homemade units do not do a proper job. They leave a smoldering mass of partly consumed carcasses and are a source of unpleasant odors.

Disposal pit. A properly constructed pit is a convenient, sanitary, and practical way to dispose of dead birds. It can be constructed rather cheaply and requires little maintenance.

The pit should be located on well-drained soil that drains away from the farm water supply. Avoid heavy clay-type soils that are poorly drained. For convenience, locate the pit as close to the poultry house as possible.

Deep burial. Direct burial is satisfactory if done properly. But the carcasses should be buried deep enough so they will not be dug up by dogs or other animals.

Commercial pick-up. This is another method of disposal. In many areas, producers do not have access to such a service. But in others, the service is provided on a regular basis. It is important to enclose dead birds put out for pick-up in a plastic bag or other fly-proof container.

Guide to disease diagnosis

Disease	Affects primarily	Cause
Laryngotracheitis *(Trachy,* LT.)	chickens, pheasants	Virus. Infected birds; unwise use of vaccine; carriers at poultry shows; airborne; contaminated clothing and equipment.
Lymphoid Leukosis *(Big liver disease)*	chickens	Virus. Egg borne or transmitted to very young chicks from infected older birds.
Marek's Disease *(Range paralysis)*	chickens	Herpes virus. Airborne or other contaminated skin and feather dust (dander). Contaminated litter. Infected birds.
Newcastle Disease	most birds	Virus. Mechanically carried; infected respiratory and digestive discharge. Contaminated equipment, shoes, clothing. Contact with infected birds.
Avian Pox	most birds	Virus. Direct contact with infected birds. Mosquitos carry virus from wild and other birds.
Paratyphoid	poults, chicks	Bacterium *Salmonella sp.* Eggshell penetration. Eating or contact with droppings of infected carriers.
Pullorum Disease *(BWD)*	chickens, turkeys pheasants, guineas	Bacterium *Salmonella pullorum.* Egg transmission from infected hen. Discharge from nose, mouth, or feces of infected chicks.
Bumblefoot	older, heavier birds	Bacterium *Staphylococcus aureus.* Cuts or bruises on foot pad allow entrance of organisms.
Leg Problems	all birds	Chemical poisons. Accidents. Inadequate nutrition. Lack of vitamins. Bacteria, virus or mycoplasma infections.
Worms	all birds	*Roundworms:* eggs directly transmitted from bird to bird through feces. *Tapeworms:* indirectly transmitted by intermediate hosts (flies, sowbugs, etc.) that contact contaminated feces.

Signs and lesions	Prevention (P) and treatment (T)
Rapid spread. Coughing, sneezing, gurgling. Blood or cheesy plug in windpipe. May be high mortality.	(P) Vaccinate, but only if a problem in your area. Do not vaccinate unnecessarily. (T) None.
Weight loss. Green droppings, tumors, enlarged liver. Sick birds usually die. Deformed, thickened legbones.	(P) Brood away from older chicks. (T) None.
Ocular: gray eye. *Skin:* enlarged feather follicles. *Nerve:* Paralysis of wings, legs, neck. *Visceral:* tumors of testes, ovary, liver, spleen, lungs, heart and proventriculus.	(P) Vaccinate, day-old chicks. Buy vaccinated chicks. (T) None.
Gasping, coughing, nasal discharge, incoordination, paralysis. Rapid spread, high mortality. Adults may show only respiratory symptoms and egg production drop.	(P) Vaccination. (T) None.
Dry pox: small yellow "warts" on wattles, comb, face. These increase in size. Dark brown scabs form, then drop off. *Wet pox:* yellow, cheesy lesions in mouth or windpipe.	(P) Vaccination is recommended in areas of large mosquito populations. (T) Swab lesions with Lugol's solution of iodine.
Huddling near heat, closed eyes, drooping wings, diarrhea, pasted vent, increased peeping sounds.	(P) Egg sanitation. Rodent and snake control. (T) Drugs, antibiotics. Follow label recommendations.
Pasted vents in chicks 1 to 21 days old, sudden death or huddling, cecal cores, pneumonia.	(P) Buy pullorum-free chicks. (T) Various drugs and antibiotics. Follow label recommendations.
Lameness, swollen foot, scab on foot pad.	(P) Avoid high roosts, sharp litter. (T) Open abcess with sharp knife, remove pus, paint with iodine or sulfa ointment.
Swollen joints, soft bones (rickets), twisted legs (perosis), broken bones, swollen feet, paralysis.	(PT) Determine causes and use the recommended treatment or preventive measures.
May cause unthriftiness and slow growth. *Roundworms:* 3 to 6 inches long, pencil lead in diameter in intestine; 1 to 2 inches long in ceca. *Tapeworms:* flat ribbon-like, segmented.	(P) Rotate birds in yards or pens. Screen off areas of heavy fecal deposition. (T) Various drugs. Follow label. Unwise use may cause more damage than the worms.

Guide to disease diagnosis

Disease	Affects primarily	Cause
Air Sac Disease Mycoplasmosis *(CRD, colds, infectious sinusitis)*	chickens, turkeys all ages	Bacterium *Mycoplasma gallisepti cum.* Egg transmitted. Also by contact with infected birds (healthy carriers).
Aspergillosis *(Brooder pneumonia)*	chicks, poults	Fungus *Aspergillus fumigatus.* Birds inhale spores from moldy feed, litter, dust.
Blackhead *(Enterophepatitis)*	turkeys	Protozoon *Histomonas meleagridis.* Droppings from infected birds or infected cecal worm eggs.
Infectious Bronchitis *(Bronk)*	chickens	Virus. Airborne from infected flocks.
Botulism *(Limberneck)*	all birds	Toxin (poison) produced by bacterium *Clostridium botulinum.* Birds eat decaying animal or vegetable material. Toxin is soluble so can be in water or maggots.
Fowl Cholera	most birds	Bacterium *Pasteurella multocida.* Contact with feces of sick birds, carcasses of dead birds. Contaminated soil, water, feed. Rodents.
Coccidiosis *(Cocci)*	chickens, turkeys (One of the most prevalent diseases worldwide.)	Protozoon *Eimeria sp.* (9 species in chickens, 7 species in turkeys.) Eating droppings containing infective parasites. Coccidia invade intestinal tract lining, produce tissue damage while multiplying.
Infectious Coryza *(Roup, colds)*	chickens only	Bacterium *Haemophilus gallinarum.* Recovered apparently healthy birds remain infected (carrier birds). Contact at poultry shows. Sick birds. Dust or water contaminated by nasal exudate.
Enteritis *(Diarrhea)*	all birds	Many causes, most unknown. High salt in feed. Droppings of infected birds.

Signs and lesions	Prevention (P) and treatment (T)
Coughing, sneezing, runny nose. Stress or secondary infection increases severity. Cloudy air sacs, often with yellow exudate. Transmitted slowly through flock.	(P) Don't mix age groups. Get chicks or poults from MG-feed birds. (T) Encourage eating. Some antibiotics help.
Gasping, loss of appetite, increased thirst. Liver swollen with circular spots.	(P) Avoid sources of mold. Control dust. (T) Clean, disinfect. Change litter.
Darkening of head, loss of appetite, droopiness, sulfur-colored droppings. Cecum has thickened wall, cheesy core.	(P) Keep turkeys away from chickens. (PT) Hepzide, Enheptin, Emytryl. Follow veterinary and label recommendations.
Rapid onset. Sneezing, coughing. watery eyes. Flock symptoms may last 10-14 days. Production drops. Small or soft-shelled eggs.	(P) Vaccinate before lay if an important problem in your flock. (T)None.
Extreme weakness. Paralysis of legs, wings, neck. Bird cannot swallow. Feathers easily removed. Head hangs. appetite. Swollen wattles, difficult breathing, dark head and wattles.	(P) Clean yards. Don't use spoiled food or scraps. (T) Place bird in shade. Fill crop with water twice daily. Give Epsom salts (1 lb. to 5 gal. water) into crop. Remove dead animals and decaying material.
Birds may die before there are visible symptoms. Dead on roost, yellow-green diarrhea.	(P) Clean ground, good management. Eliminate rodents, raccoons, opossums, etc. (T) Sulfaquinoxaline, except in laying hens. Use cleared antibiotics. Complete clean out.
Possible high mortality. *Intestinal:* Pale, droopy, huddled, use less feed, water. Production drops. *Cecal:* Bloody droppings, cecal blood. In turkeys, cheesy ceca.	(P) Use preventive drugs (coccidiostats). Screen droppings from birds. (T) In acute outbreak give recommended drugs in water. Use according to directions.
Rapid onset. Svvollen sinuses, nasal discharge, eyelid may stick shut, drop in feed consumption and egg production.	(P) Don't mix age groups. (T) Antibiotics or sulfa drugs effective in some cases.
Watery, discolored droppings. Layers may drop in production.	(P) Sanitation and good ration. (T) Get specific diagnosis. Antibiotics. Copper sulfate may help. For a stock solution use 1 lb. $CuSO_4$ and 1 to 2 cups vinegar per gal. of water. Use 1 tbsp. of stock solution to 1 gal. drinking water. Avoid metal waterers.

Chapter 13
Raising ducks, geese, turkeys, bantams, and guinea fowl

CHICKENS ARE ONLY ONE OF several poultry enterprises available to the family that wants to raise a small flock. Ducks, geese, turkeys, and guinea fowl are others often raised on small acreage. There also are bantam chickens, which are similar to those described earlier, but smaller and with different requirements.

All of these birds have the same general requirements. First, you must conform with zoning regulations in your home areas. Secondly, they all require daily care. That means that someone in the family must take the responsibility for daily chores.

The economic returns are not as good as for a flock of layers or fast-growing broilers. However, each type of bird does have somc advantages.

Millions of ducks, geese, and turkeys are raised each year. Most are produced in commercial enterprises and are raised by the thousands. However, you can raise your own for family use or for limited sale.

Ask yourself these questions: How many ducks or geese would our family eat in a year? How many turkeys? Is one of these enterprises worth the effort?

Much like chickens

The incubating, brooding, and rearing of these birds is much like that of chickens. Refer to the chapters on these subjects that appear in the front of this book. Only minor adaptations are needed for ducks, geese, turkeys, and guinea fowl.

How to raise ducks

Around 10 million ducks are raised annually for meat in the United States. Most are produced under confinement on specialized duck farms in a few commercially important duck production areas. Many farms still raise a few ducks primarily for family use or local sale.

Family flock ducks are raised primarily for meat. Although most breeds used are relatively poor layers, the flock should be managed to save the eggs produced for food purposes or hatching. The commercial duck industry is built around the Pekin breed.

Pekins reach market weight early and are fairly good egg producers, but are poor setters and seldom raise a brood. The Rouen is a popular farm flock breed. It is slower growing than the Pekin, but reaches the same weight over the five- to six-month period of feeding and foraging under farm flock conditions. Its slower growth and colored plumage make it undesirable for mass commercial production. The Muscovy, a breed unrelated to other domestic ducks, is also used to some extent in farm flocks. They are good foragers and make good setters. Muscovy males are much larger than the females at market age. Meat production is generally of primary importance in selecting a breed, but egg production for propagation, brooding tendency, and the white plumage that produces an attractive dressed carcass should also be considered.

The keeping of small, ornamental varieties of ducks, sometimes called bantam ducks, for exhibition or hobby purposes is increasing. Included in this grouping are White and Gray Calls, Black East Indias, Wood Ducks, Mandarins, and sometimes Teal. Most general poultry shows and some bantam shows offer classes for these ducks.

Natural incubation

Natural methods of incubation are frequently used on small farms. This is especially true when Muscovies are raised. These ducks are excellent setters and will incubate their eggs easily.

Other meat- and egg-type breeds do not set regularly. These breeds can be naturally hatched under broody hens.

Clean, dry nesting facilities must be provided for setting hens and ducks. Feed and water should be within close proximity because the female must obtain her daily requirements within short periods of time. Delay in finding feed and water will result in undue chilling of the eggs. This situation is more critical with broody hens because they will have to keep their nests for one week longer than if they were setting on chicken eggs: Muscovy eggs require 35 days of incubation; all other domestic duck eggs require 28 days of incubation.

Artificial incubation

Incubators designed for hatching duck eggs are available, but when only a few eggs are to be hatched the regular chicken-egg machine may be used. For best results, follow the manufacturer's directions.

Start the incubator a day or two in advance of the first setting. This will allow you to bring the machine into correct incubation condition before the eggs are set.

Place the eggs in the incubator small end down. In some small machines, eggs can be incubated level.

Most incubators are equipped with automatic turning devices. Set the device to turn the eggs every three hours. If manual turning is necessary, it should be done at least three or four times daily.

Candle the eggs after seven to eight days of incubation. This can be done by passing each egg over a small hand candler or by placing an entire tray of eggs above a bright light.

Remove infertile eggs and eggs containing dead embryos from the incubator. Cracked eggs and those with ruptured yolks can also be detected and removed during the candling operation.

Eggs are frequently candled again after 25 days of incubation (32 days for Muscovy eggs). At this time the bills of normally developing ducklings can be seen within the air cells. Considerable movement can also be seen.

Ducklings can be artificially brooded in about the same way as baby chicks. Due to their rapid growth, ducklings will need heat a shorter period of time and floor space requirements will increase more rapidly.

Any small building or garage or barn corner can be used as a brooding area for small numbers of birds. The brooding area should be dry, reasonably well-lighted and ventilated, and free from drafts. Cover the floor with about four inches of absorbent litter material, such as wood shavings, chopped straw, or peat moss. Litter dampness is more of a problem with ducks than with chicks. Good litter management will require removal of wet spots and frequent addition of clean, dry litter. Be sure litter is free of mold.

Infrared heat lamps are a convenient source of heat for brooding small numbers of birds. Use one 250-watt lamp for 30 ducklings. Heat lamps provide radiant heat to the birds under them. Since the air isn't heated, room temperature measurement isn't so important.

When using hover-type brooders, brood only half as many ducklings as the rated chick capacity. Because ducklings are larger than chicks in size, it may be necessary to raise the hover three to four inches higher. Have the temperature at the edge of the hover 85° to 90° F. when the ducklings arrive. Reduce it 5° to 10° per week.

Confine the birds to the heated area with a corrugated paper chick guard for the first three to four days. Watch the actions of the birds as a clue to their comfort. If they are too hot, they will move away from the heat. If too cold, they may pile up and be noisy.

High temperatures may result in slower feathering and growth. Supplementary heat may be needed for five to six weeks in cold weather; in summer, only two to three weeks. By four weeks of age the ducklings should be feathered enough to be outdoors except in extremely cold, wet weather. In some areas attention to predator control may be necessary when the ducklings are turned out.

Allow one-half square foot of floor space per bird during the first two weeks. Increase this to at least one square foot by four weeks. If the birds are to remain confined after the first month, provide them with at least two square feet of floor space.

Start feed early

Ducklings should have feed and drinking water available when they are started under the brooder or hen. Use waterers the birds can't get into. This is especially important in the brooding area since ducklings are easily chilled when they become wet while still in the "down" stage. Pans or troughs with wire guards are satisfactory. Place waterers over low, wire-covered frames to help reduce wet litter problems. Change waterers or adjust size as birds grow. The waterer should be wide enough and deep enough for a bird to dip its bill and head.

Nutrition for ducks

Maximum efficiency for growth and reproduction can be obtained by using commercially prepared diets, which are sold in pellet form. Pellets are recommended instead of mash because they are easier to consume, they reduce waste, they do not blow away, and feed conversion is usually superior. Satisfactory results are possible with mash.

Four diets are recommended: Starter, grower, breeder-developer, and breeder. Use pellets ⅛-inch in diameter for the starter diet; use pellets 3/16-inch in diameter for the grower, breeder-developer, and breeder diets.

Feed ducklings the starter diet the first two weeks after they hatch. To encourage early consumption, place the feed in baby-chick-sized hoppers and locate them close to the water supply. When ducklings reach two weeks of age, switch them to the grower diet. Feed this diet until ducklings are marketed.

Young drakes and ducks selected as potential breeders should be fed a breeder-developer diet. This special diet contains less energy per pound of feed than the starter or grower diet. When fed in restricted amounts, the diet will keep the breeders from putting on excess fat but it will provide the nutrients needed. For each 10 breeders feed 4 ½ pounds of the breeder-developer diet daily. Feed half in the morning and half in the late afternoon.

Increased requirements for reproduction make it essential to feed breeding stock a breeder diet. Switch the breeders to the breeder diet about one month prior to the date of anticipated egg production. To insure good eggshell quality, give the breeders oystershells. They can be fed free choice in separate hoppers within the breeder pens.

In some areas commercial suppliers have feeds formulated for duck feeding. Check with the suppliers in your vicinity. Growers desiring to mix their own feeds should write to their state university extension poultry specialist for formulas. If duck feeds aren't available, start ducklings on crumbled or pelleted chick starter for the first two weeks. Place feed for the first few days on egg case flats or other rough paper; slick-surfaced paper may cause leg injuries. Ducklings can be fed a pelleted grower ration plus cracked corn or other grain. Keep feed before the birds at all times and provide insoluble grit.

Ducks are easy to raise because they are hardy and not susceptible to many of the common poultry diseases. The use of medicated feeds isn't usually necessary. Very few additives have been approved for nutritional or medicinal use in duck feeds. Waterfowl may be more sensitive to some drugs than other poultry. Incorrect use of certain medicated feeds formulated for chickens and turkeys could harm ducklings.

Small flocks of ducklings raised in the late spring with access to green feed outdoors generally have few nutritional problems. While ducks are not as good foragers as geese, they do eat some green feed and farm flocks are usually allowed to run at large. Cut green feed can be supplied to the birds when they must be kept inside in bad weather. Water for swimming isn't necessary for successful duck production.

Under commercial conditions, Pekin ducklings are ready for market when seven to nine weeks old. These birds weigh 6 to 7 pounds and have consumed 20 to 25 pounds of feed. Rouens raised under farm flock conditions may take five months to reach these weights. Muscovy ducks take somewhat longer. The holiday retail duck market is greatest from Thanksgiving through New Year's Day. Ducks grown for home use or limited local sales can be slaughtered any time. If ducklings are kept longer than eleven to twelve weeks, new pinfeathers begin to come out, making it difficult to pick them clean for another several weeks.

Breeder flock management

Select stock from flocks hatched in April and May. Using males from early flocks will help insure their readiness for mating for the start of the following year. Choose vigorous birds with good weight, conformation, and feathering prior to marketing the young flock. Keep one male for each five to six females. Young birds should be selected only from families that have good egg production, hatchability, and fertility records.

Identification of males and females is necessary when selecting birds for breeder flocks and for exhibition. Even in breeds that have a differentiated color pattern, both sexes may resemble each other in their summer plumage. Ducks and geese can be sexed by everting the vent and examining the reproductive organs. This practice requires some experience and may be more easily done with day-old birds or during the breeding season. In some breeds mature males develop characteristic curled feathers at the base of the tail. After about six weeks of age the sounds ducks make can be a clue to their sex. Females have a more definite sharp quack, while males have a sound that is not nearly so loud or harsh but more of a muffled sound.

Birds held for breeders must be kept from becoming too fat. The breeder-developer ration fed during the holding period should contain less energy than starter and grower rations. If the grower ration is continued during the holding period, gradually restrict feed to about 70 percent of the amount fed at the start.

Change to a breeder-laying ration about one month before egg production starts. Don't bring birds into production before seven months of age. Feeding oystershell is optional to improve eggshell quality. Increasing day length with lighting stimulates egg production. Provide a 14-hour day three weeks before the desired egg production date. The flock should be laying at a high rate of production within five to six weeks. Meat-type breeds should remain above 50 percent for about five months. Breeders should be provided with a clean, dry, well-ventilated shed or house. Allow five to six feet of floor space per bird. Birds are often confined at night to get a maximum number of eggs and then allowed day-time access to the yard. Provide floor level nest boxes. Most eggs are laid in early morning. Gather eggs about 7 a.m. and let the birds out of the house. If some birds stay on the nests, a second collection can be made later. Clean, dry litter and nesting material will help produce clean eggs.

Soiled eggs should be cleaned soon after gathering. They should be washed in warm water (at least 20° warmer than the eggs) containing an egg cleaning and sanitizing compound used in accordance with the instructions of the manufacturer. Store eggs for hatching at 55° F and a relative humidity of 75 percent. Eggs stored longer than two weeks may decline in hatchability. If stored more than a week, turn eggs daily to prevent yolks from sticking to shells.

Diseases of ducks

Ducks raised in small numbers and in relative isolation suffer little from diseases. But where large numbers of ducks are concentrated, diseases may be extremely widespread. Muscovies appear to be resistant to diseases common to Pekin and Runner-type ducks.

How to raise geese

Geese make a good family project. Remember, they are big birds and take more space. And since they make good use of green feed, you need a few acres for them.

Let's start with some little-known facts: 1) Geese are easy to raise, 2) they grow rapidly, 3) they do not require much expensive feed, and 4) they are highly disease resistant. How is that for a few good points for starters? Geese can create one problem if allowed to do so by their owners: They can be dirty if confined to a very small area.

Which breed of geese?

What size of goose do you want? There are five breeds that are sometimes available. The table below shows size and color as well as the names of the five major breeds.

Probably the most popular breeds for marketing are the Toulouse and the Embden. Many African and White Chinese are also raised. If you wish to raise the birds as a hobby, the choice of breed is really one of your fancy. If you are raising for a market, you are raising for someone else's fancy. Generally, people who buy geese for the consumers market want birds that are large, young, and generally white-feathered.

Common Breeds of Geese			
Breed	**Adult weight (lbs.)**		**Color**
	Male	**Female**	
Toulouse	26	20	Dark gray with white abdomen. Dewlap present
Embden	26	20	White
African	20	18	Gray with brown shade (has a knob on its beak)
Pilgrim	14	13	Male white, female gray and white
Chinese	12	10	White or brown (has knob on its beak)

Where are they?

Very little research has been done with geese. Almost no selective breeding has been done with them. Most of the geese in the United States are raised in the west north central states. According to the past U.S. census reports, Wisconsin, Minnesota, Iowa, and North and South Dakota have been leaders (with much fluctuation) in goose production with Washington, California, Indiana, Ohio, and Pennsylvania also raising and selling some geese.

If you want to buy a few goslings, your local county extension office may know of someone in your area who keeps geese. These growers may have goslings to sell or may know where you can obtain them.

Incubation of goose eggs

The incubation period for eggs of most breeds of family flock geese varies from 29 to 31 days. Four to six eggs can be incubated under a setting hen and ten to twelve under a goose. Mark the eggs so that they can be turned by hand twice daily if the setting hen does not turn them. Better hatchability is reported by some growers if the eggs are lightly sprinkled or dipped in lukewarm water for half a minute daily during the last half of the incubation period. Remove goslings from the nest as they hatch, and keep them in a warm place until the youngest are several hours old. Since it takes time to complete the hatch, if the goslings aren't removed as hatched, the hen may desert the nest, leaving with the hatched goslings before the hatch is completed.

Goose eggs can be hatched in either still-air or forced-draft incubators. Follow the instructions from the machine's manufacturer. You may increase the success of your hatching operation if you can talk with a person who has had success with machine incubation of goose eggs.

When to start goslings

First of all, order goslings to arrive in late May or early June. The birds require much less care at that time of year. They do not have to be protected from so much cold for such a long period of time. Goslings require a warm, dry place for one, two, three, or four weeks, depending on the weather.

Facilities for brooding

A special building is not required for brooding small numbers of geese. Any small building or a corner of a garage or barn can be used as a brooding area for a small flock if it is dry, reasonably well-lighted and ventilated and free from drafts. Cover the floor with four inches of absorbent litter material such as wood shavings, chopped straw, or peat moss. Maintaining good litter requires frequent stirring, removal of wet spots, and periodic addition of clean, dry litter. Be sure litter is free from mold.

Heat lamps are a convenient source of radiant heat for brooding small flocks. Use one 250-watt lamp for each 25 goslings.

When using hover-type brooders, brood only about one-third as many goslings as the rated chick capacity of the hover. Because goslings are larger in size, with some brooders it may be necessary to raise the hover three to four inches higher than for baby chicks. Have the temperature at the edge of the hover 85° to 90° F. when the goslings arrive. Reduce the temperature 5° to 10° per week until 70° F. is reached.

Confine the birds to the heated area for the first three or four days with a corrugated paper or wire mesh fence. The behavior of the birds is an indication of their comfort; they will move away from the heat source if they are too warm or may crowd together at corners or under the brooder if too cold. If there is no light at the heat source, a dim light on the birds at night tends to discourage crowding.

High temperatures may result in slower feathering and growth. Heat is usually not needed after the fifth or sixth week, and in good weather, the young geese can be placed on pasture. In warm weather, goslings can be let outdoors even at two weeks, but must be sheltered from rain. They must be kept dry to prevent losses from crowding and chilling while in the "down" stage.

Allow at least one-half square foot of floor space per bird at the start of the brooding period. Increase this to one square foot by two weeks. If birds are to remain confined due to inclement weather, be sure additional floor space is provided as they increase in size.

Feeding your geese

Goslings should have feed and drinking water when they are started under the brooder or hen. Use waterers the brooders can't get into to prevent losses from chilling. Waterers should be wide and deep enough for the bird to dip both bill and head. Pans or troughs with wire guards are satisfactory. They should be placed over screened platforms to aid in keeping litter dry. Change waterers or adjust size as birds grow.

Feeds formulated for goose feeding programs are not normally available from commercial suppliers. Growers who wish to mix their own feeds should contact their extension poultry specialists for formulas.

Goslings can be started on a crumbled or pelleted chick starter. Place feed the first few days on egg case flats or other rough paper. Use the same type of feeders as used for chicks, changing type or adjusting size as the birds grow. Keep feed before the birds at all times and provide insoluble grit. After the first two to three weeks, a pelleted chick grower ration can be fed, supplemented with cracked grain.

Geese are quite hardy and not susceptible to many of the common poultry diseases, so medicated feed is not generally necessary. Certain coccidiostats used in starting and growing mashes may cause lameness or even death in goslings.

Geese are excellent foragers. Good succulent pasture or lawn clippings can be provided as early as the first week. By the time the birds are five to six weeks old, a good share of their feed can be from forage. Geese can be very selective and tend to pick out the palatable forages. They will reject alfalfa and narrow-leaved tough grasses, and select more succulent clovers, bluegrass, orchard grass, timothy, and bromegrass. Geese can't be grown satisfactorily on dried-out, mature pasture. Corn or pea silage can be fed if available.

An acre of pasture will support 20 to 40 birds, depending on the size of the geese and pasture quality. A three-foot woven wire fence will ordinarily confine the geese to the grazing area. Be sure that the pasture areas and green feed have not had any chemical treatment that may be harmful to the birds. They should be provided with shade in hot weather.

Although supplemental grain feeding of goslings is often continued after they have been established on good pasture, many flocks are raised on green feed alone during the pasture period. Geese to be marketed should be fed a turkey finishing or similar ration for three to four weeks before processing. Any birds saved for breeding stock should not be fattened.

Farm geese are usually sold in time for the holiday market in late fall when they are five to six months old. They will weigh from 11 to 15 pounds depending on the strain and breed. Some young geese (also called green geese or junior geese) full-fed for rapid growth are also marketed at 10 or 12 pounds when they are 10 to 13 weeks old. For several weeks after this age geese have many pinfeathers that are difficult to remove during processing. Growth of geese after 10 to 13 weeks is very slow, compared with the rapid growth of the young gosling.

Considerable attention has been given to the use of geese to control weeds in crops and gardens. Development of more selective herbicides is reducing this practice. The problems in coordination of bird supply and management with weed and crop growth makes goose weeding rather impractical for most producers.

Need-only shelter

Geese do not do well if enclosed in a house. They should be confined to a yard with a house for protection during winter storms. When green pasture is not available, breeders can be maintained on roughage such as leafy clover or alfalfa hay, corn or pea silage, with a small amount of grain. If breeding stock becomes overly fat, poor fertility and hatchability may result.

Start feeding a pelleted breeder ration at least a month before egg production is desired. Provide water at all times, as well as a supply of oystershell and grit. Lights in the breeder house can be used to stimulate earlier egg production if desired. Geese start laying in February or March and often lay until early summer. Nest boxes should be provided to aid in obtaining a maximum number of eggs and reduce the amount of cleaning required. Boxes should be at least two feet square and built with partitions or spread some distance apart to reduce fighting. Large boxes or barrels are frequently used for range nests. The heavier geese lay from 20 to 50 eggs per season, depending on the amount of selection for egg production in the strain being raised.

Eggs should be gathered twice daily, especially during cold weather. They should be stored at 55° F. and a relative humidity of 75 percent until set for hatching. Eggs should not be held for more than seven to ten days, and should be turned daily if kept more than a few days.

Breeder flock management

Select geese that are vigorous and well-developed, have shown rapid growth, and have compact, meaty bodies. Matings should be made at least one month prior to the breeding season (around the first of the year). The larger breeds of geese usually mate best in pairs and trios. Ganders of some lighter breeds will mate satisfactorily with four or five females. Males will usually mate with the same females year after year.

Sexing geese

It is difficult to distinguish sex in geese except by examination of the reproductive organs. Lift the goose by the neck and lay it on its back, either on a table or over your bended knee, with the tail pointed away from you. Move the tail end of the bird out over the edge so it can be readily bent downwards. Then insert your pointer finger (sometimes it helps to have a little Vaseline on it) into the cloaca about half an inch and move it around in a circular manner several times to enlarge and relax the sphincter muscle that closes the opening. Next, apply some pressure directly below and on the sides of the vent to evert or expose the sex organs.

In some birds the male organ is somewhat difficult to unsheath. An inexperienced sexer may easily call a bird a female if, after slight pressure, the corkscrew-like male organ is not exposed. However, only the presence of a female genital eminence will positively identify a female.

Selection and care of bantams

The bantam is a miniature fowl available in over 150 breeds and varieties of domestic chickens. History indicates they came from Asia. Bantams are hardy, inexpensive birds to raise and maintain. Bred for beauty and form, they can be reasonable egg layers and meat producers. Egg size falls in the peewee and pullet-weight classes and recipes calling for two large eggs would require three bantam eggs.

Characteristics of bantams

Size. Generally adult bantams weigh between one and two pounds. Often, with the introduction of new bloodlines and using growth-promoting diets, bantams tend to increase in size with each generation. Continued selection for smallness is necessary.

Growth rate. The growth habits of bantams vary with each breed. Usually, the tight-feathered breeds mature in about six months, while the loose-feathered breeds take about nine months. Bantams tend to slowly gain size and weight through their second year.

Meat and egg production. Bantams have not been selected or bred especially for meat or egg production. The meat is delicious, but the portions are small. The eggshell color varies with the breed and ranges from white for Mediterranean breeds to light brown for miniatures of the Asiatic breeds. Aracuanas are an exception, with eggshell color in the blue to green range.

Broodiness. Broodiness is the inherent tendency of the hen to remain on the nest and incubate eggs. Most bantam hens will go broody, but the Mediterranean varieties are poor risks, as they frequently desert their nests. Miniatures of the heavy breeds (especially the feather-legged ones) are exceptionally good brooders, mothers, and foster mothers for other breeds.

Breeds and varieties. Bantams are identified by breed the same as other fowl and livestock. Breed differences are those of shape and body outline. Variety difference usually consists of differences in color patterns or comb type. The selection available is tremendous.

- It is easier to produce good specimens of solid-color varieties.
- Games and Cornish need plenty of exercise to develop desired hard feathering and muscle tone.
- Feather-legged breeds should be kept out of wet grass and muddy areas.
- Crested breeds, such as Polish and Houdans, should be kept dry in cold weather.
- Adult modern and Old English Game males must be dubbed—comb and wattles removed.

Selecting and mating

Start off with good, vigorous birds. Select your breeders carefully, using only your best stock. Pen these birds together (one cock and two to eight laying hens) ten days before starting to save the eggs. Trim the fluff feathers from around the vent to insure better fertility of the birds that are heavily feathered. Other, less than perfect, hens may be used for brooding.

Bantam eggs require 21 days to incubate. You may set the eggs under a broody hen, build or purchase a small incubator of your own, or have your eggs custom hatched.

If you have identified the eggs by pen matings, then you must set them under separate hens or in separate wire boxes in the incubator. Use a toe web punch or leg bands to mark the chicks.

Setting the brooding hen

- Select a quiet, secluded area protected from all predators including rats.
- Provide a darkened, easy-to-enter nest at ground level.
- If several brooding hens are setting in the same area, separate so they cannot trade nests. Dust hens with poultry louse powder to be sure that no parasites exist.
- Give each hen only the number of eggs she can completely cover (eight to ten).
- Provide fresh feed and water daily. Feed mostly grain to the brooding hen, as mash will tend to make her droppings loose and she may soil her eggs.
- Leave her alone.

After the eggs hatch, place hen and chicks in a light, dry, protected area with feed and water. After one week, allow outdoors in movable wire pen with rain protector.

Artificial brooding temperature starts at 92° F. and drops 5° each week. Place cloth over litter for first two weeks. Change daily. Newspapers are slick and may cause crippled legs. Do *not* attempt to raise breeds of different sizes or chicks of different ages together. Feather picking and cannibalism may start.

Feeding bantams

Baby chicks have a three-day reserve supply of yolk, but should be fed the day after hatching. They should receive an 18 percent to 20 percent chick starter mash or crumble containing a coccidiostat, and have a constant supply of clean water available. At six weeks of age, add small amounts of chick scratch daily. Increase the scratch approximately 5 percent each week, until by 16 weeks of age, they are receiving half scratch and half growing mash.

Switch to growing mash for growing birds or 15 percent protein, multipurpose mash at eight weeks of age. Continue with growing mash and scratch mixture until pullets start to lay.

Laying birds should have laying mash plus a small amount of grain for the activity of scratching. Birds in molt and roosters do well on half scratch and half laying mash. Supply grit or small stones for the gizzard every month or so. Supply oystershells continually if the birds are running loose or being fed other than an all-mash diet.

Housing for bantams

As bantams are usually kept as a fun project, make the house large enough or accessible enough that you can enjoy it, too. The coop should be draft-free, insulated, and easy to ventilate. Under extreme winter conditions, it might be good to provide some source of heat to prevent frozen combs and toes.

Allow two square feet of floor space for large varieties and 1 ½ for smaller varieties. Six inches of litter is recommended for floor birds. Some birds, such as the feather-legged varieties, do well on wire floors that raise them above the droppings. Provide one nest for every four or five hens and keep the nest well supplied with a deep nesting material.

Growing a small flock of turkeys

Turkeys can be raised successfully on small farms, but they require special care and equipment. Young turkey poults must be kept warm and dry.

Raise turkeys away from chickens and other birds in order to prevent disease. Sinusitis and blackhead can be serious problems when turkeys and chickens are raised together.

The most common variety of turkey is the Large White. Similar in size, but less available, is the Broad Breasted Bronze. With good management, roaster-size turkeys can be produced efficiently using either of these varieties. Hens will grow to live weights of about 15 pounds at 22 weeks of age; toms will weigh about 25 pounds at 24 weeks of age.

The Beltsville Small White also satisfactorily produces small roaster-size turkeys—at 22 weeks hens average 8.7 pounds and toms average 13.6 pounds. Excellent fryer-roasters are produced with Beltsville Small Whites when they are marketed at about 16 weeks. Large White females, marketed at 13 weeks, make satisfactory fryer-roasters.

Flocks are usually started with day-old poults purchased from a hatchery or feed store. The poults should come from sources free of pullorum, sinusitis, and other diseases. Start with a healthy flock. The operation of a brooder is practical if you buy 20 or more poults. If you want to raise fewer than 20, try to purchase turkeys that are six to eight weeks old, from a commercial grower, when they no longer require brooding.

Practical production standards

	Hens	Toms
Age at slaughter	20 weeks	22 weeks
Total feed consumed	60 lbs.	80 lbs.
Average live weight	16 lbs.	27 lbs.
Shrink from live weight to oven-ready weight	21%	21%
Flock mortality to slaughter age	10%	10%

Housing and equipment

Poults need a dependable source of artificial heat during the first few weeks of life. The first week, provide uniform temperatures of about 95° F. Thereafter, lower the temperature 5° per week. For a small group of poults, use an infra-red bulb, an electric coil, a shielded electric light, or the heating element of an outdoor (sunshine) wire-floored brooder.

To brood 100 or more poults, you might consider a commercially-used gas or electric brooder that has a hover and automatic temperature controls. During the first week, confine the poults to the heated area containing feed and water. Corrugated cardboard, at least 12 inches high and available in rolls, is often placed in a circle about three feet from the hover to confine poults during this period. In hot weather a guard may be made of poultry netting. Check temperature closely, especially the first week, adjusting it if the poults show signs of distress. Although heat may be discontinued at four weeks of age in warm weather, provide heat until six or seven weeks of age in cool weather.

Poults on litter need at least 1 ½ square feet of floor space per poult to six weeks of age. When brooding poults on the floor, you can use ⅛ inch of sand for litter the first two weeks; then add one inch of chopped straw or shavings. During hatching and brooding, avoid slick surfaces (such as newspaper), which can cause serious leg problems. Be sure to protect the poults from cats, dogs, and other predators, both during and after the brooding period.

After brooding the poults, place them in a yard, allowing at least 30 square feet of yard space per turkey. A four-foot fence usually will confine turkeys of the heavy varieties. Avoid standing water by selecting a well-drained site. Roosts are generally unnecessary but, if used, can be made of two-by-fours laid flat 20 to 24 inches apart and supported about 15 inches off the ground. Allow 8 to 12 inches of roost space per bird. Select feeding and watering equipment that is easy to use and remains as clean as possible. Provide the turkeys with a shelter to protect them from sun and rain—100 to 180 square feet of roof per 100 turkeys, seven to eight feet off the ground.

Temperatures of 100° F. or more can cause mortality, particularly when there is insufficient shade or water. On extremely hot days, when temperatures are 105° F. or above, either wet the turkeys hourly with a hose, or use foggers. Mature and near-mature turkeys do not tolerate high temperatures as well as young turkeys do.

Feeding and management

Feed and water the poults as soon as possible after bringing them home, dipping their beaks in water to help them learn to drink. The first two days, place some feed in paper plates or on the lid of the poult box. Feed and water should be available at all times.

Provide a one-gallon drinking fountain for each 75 poults. Allow 0.9 linear inches of water trough space per turkey after eight weeks of age. Water may be piped to the turkey pen and controlled with a float valve, or commercially available waterers may be used. Provide a four-foot feeder, or its equivalent, per 100 poults (one inch per turkey) to three weeks, two inches per turkey to eight weeks, and three inches per turkey thereafter. With circular feeders, less feed space is needed. For example, a range feeder with a 300-pound feed capacity is adequate for 75 turkeys after they reach eight weeks of age. It takes about 70 pounds of feed to raise the average large turkey to market age.

Since turkeys are fast growing, it is very important to buy correctly formulated turkey feeds. If the feed manufacturer's directions are available, follow them. Percentage of protein in the total ration required by turkeys decreases as birds grow older.

For the first four weeks, when feed intake is low, poults need a turkey starting mash containing 28 percent protein. After four weeks, turkeys can be fed a less expensive mash with lower protein content. At eight weeks of age, a grow mash with 22 percent protein is recommended. Feed with less and less protein can be used thereafter. Whole or cracked grain such as milo, which is commonly used in many areas, can be fed to turkeys over 10 or 12 weeks of age. Equal parts of growing mash or pellets (22 percent protein) provide an overall ration containing about 16 percent protein. Turkeys fed on whole grains should receive granite grit also. Hen turkeys in lay should have access to a feeder containing oystershell or limestone grit.

Preventing disease problems

To a large extent, the health of your flock will depend upon sanitation and avoiding contact with other flocks of turkeys and birds of all kinds. Clean feeders and waterers help prevent coccidiosis and infestations of blackhead and roundworm. When the poults are young, wash the waterer at least once a day and supply clean water. At the same time remove any droppings or litter that may be in the feeder. Never use moldy feed or litter. Molds may cause serious respiratory or intestinal problems. Placing feeders and waterers on wire platforms reduces the chance of disease by making it easier to keep them clean and by preventing turkeys from picking at nearby droppings. As far as possible, exclude from your turkey pen visitors who have contact with poultry and all kinds of birds, especially chickens and upland game birds.

Take precautions to avoid cannibalism. If you don't have a debeaker, borrow one from your feed store or from a friend. A mild debeaking, removing about one-third of the upper beak, can be performed easily when your turkeys are about three weeks old. Cannibalism is more likely to occur in older turkeys. Hens nearing maturity sometimes try to pick toms, or hens may be injured during mating. When raising turkeys for meat, it is advisable to keep the sexes separate after four months of age. Desnooding (or removing the snood) day-old poults at the hatchery is a common commercial practice that makes fighting and resulting injury less likely.

Keep away from chickens

Raising turkeys and chickens together may lead to trouble with sinusitis and blackhead. Chickens may harbor the causative agents of sinusitis or blackhead without appearing sick. Sinusitis is caused by a small, bacteria-like organism called *Mycoplasma gallisepticum.* Blackhead is caused by a microscopic protozoan parasite called *Histomonas meleagridis.* It causes pathological changes in the intestinal tract and liver and, if uncontrolled, can cause high mortality in turkeys. If your premises are heavily contaminated with blackhead parasites, consult your veterinarian or a veterinarian at a diagnostic lab about preventive feed or water medications. Medications are also available for treatment if the disease breaks out.

If your flock becomes sick, an accurate diagnosis should be obtained. An infectious disease may be involved, or perhaps the problem is poor nutrition or management. It is important to know what is wrong with the turkeys in order to treat them properly and manage the flock to prevent further losses. Be sure to take typically sick or fresh, dead birds to a laboratory and provide an accurate history on the course of the sickness.

Marketing turkeys

If you have raised more turkeys than you need, you might consider selling the extra turkeys, either alive or processed. Sometimes it is most convenient to have them custom processed at a small poultry processing plant. Although state and federal laws regulate the processing of poultry for sale, limited direct sales of home-grown and home-processed turkeys may be exempt. For details, contact city or county health departments or your state Department of Food and Agriculture, Bureau of Meat Inspection.

Home processing

Remove feed eight to twelve hours before killing, but allow turkeys access to water. Suspend the turkey by its legs held by a light rope or metal shackle. With the head held in one hand, make a deep cut across the throat from the outside, severing the jugular vein. When bleeding and movement cease, submerge the turkey in hot water to loosen the feathers. The subscald method is the most common: Dip the turkey in water at 140° F. for 30 to 40 seconds. Now rehang the turkey and remove the feathers.

After picking the turkey, wash the skin with clean water. Remove the legs at the hock joint and then remove the head. Eviscerate the turkey, saving the gizzard, liver, and heart. Finally, rinse the body cavity with clean water.

Chill the turkey by placing it in ice water for several hours. After chilling, drain the turkey and put it in a plastic bag. Either freeze the processed turkey or keep it at refrigerator temperature, preferably not more than four days. After washing up, it is a good idea to sanitize the tables and other items used in processing with diluted household bleach.

Raising guinea fowl

Many hotels and restaurants in large cities serve guineas at banquets and club dinners as a special delicacy. Prime young guineas are used as a substitute for game birds such as grouse, partridge, quail, and pheasant.

The flesh of young guineas is tender and has a fine flavor resembling that of wild game. Old guineas may be tough, and the flesh rather dry. Dressed guineas are attractive in appearance, although darker than common fowls.

Production

Some guineas are raised in flocks of 100 or more, but most are raised in smaller flocks. Through lack of good management, many farmers who keep small flocks of guineas obtain only a few young birds from each hen.

Guineas often are useful in protecting the farm flock from marauders by their loud, harsh cry and pugnacious disposition. They destroy insects in the garden. They do not scratch, and therefore are less destructive than chickens.

The majority of guineas are produced on general farms, put through the poultry-processing companies, and then shipped to city markets.

Baby-keet production

The sale of guinea hatching eggs, guinea chicks (baby keets), and guinea fowl for breeding is very limited. Only a few hatcheries have taken up baby-keet production. These hatcheries have created an interest in guineas and have provided a market for hatching eggs.

Domesticated varieties

Guineas have been domesticated for many centuries; they were raised as table birds by the ancient Greeks and Romans. They were brought to this country by the early settlers.

There are three principal varieties of the domesticated guinea fowl in the United States—Pearl, White, and Lavender. The Pearl and the White are the most highly prized.

The Pearl variety has a purplish-gray plumage, regularly dotted or "pearled" with white. It is so handsome that the feathers often are used for ornamental purposes.

The White variety has pure white plumage. Its skin is lighter in color than that of the Pearl variety.

Lavender guineas resemble those of the Pearl variety, but their plumage is light gray or lavender dotted with white.

By crossing the Pearl or Lavender varieties with the White, what is known as the "splashed" guinea is produced. Its breast and flight feathers are white, and the rest of the plumage is pearl or lavender.

Keets are very attractive. Those of the Pearl variety resemble young quail. They are brown, the under part of the body is lighter than the rest, and the beak and shanks are red. The first feathers, which are brown, are replaced gradually by the "pearled" feathers; when the birds are about two months old, the brown feathers have disappeared completely. About this time, the wattles and helmet begin to make an appearance. At maturity, birds weigh from three to three and one-half pounds.

Management of breeding stock

In their wild state, guinea fowl mate in pairs. This tendency prevails also among domesticated guineas if males and females in the flock are equal in number. As the breeding season approaches, mated pairs range off in the fields in search of hidden nesting places in which it is difficult to find the eggs.

Under domestic conditions, it is not necessary to mate the birds in pairs to obtain fertile eggs. On most general farms, one male is kept for every four or five females. When guineas are kept closely confined, one male may be mated with six to eight females, and several hens will use the same nest.

On some farms the breeders are kept confined during the laying period in houses equipped with wire-floored sun porches. Open-front poultry houses that have plenty of ventilation are desirable.

Guinea breeding stock usually are allowed free range. They are difficult to confine in open poultry yards unless their wings are pinioned or one wing is clipped. Birds on range also may be treated in this manner.

Guinea keets may be pinioned after they are from one to two weeks old by snipping off the last joint of the wing. It is more difficult to pinion an adult bird. When this operation is performed on a full-grown bird, the wing must be tied up to prevent excessive bleeding. Another method of treating adult stock is to clip the flight feathers of one wing; this should be done every year.

Feeding breeding stock

Young guineas raised for breeding should have a growing diet in fall and winter prior to egg production, a breeder diet during the laying season, and a maintenance ration after the hens are through laying.

A breeder mash containing 22 to 24 percent protein should be kept available to the birds, beginning about a month before eggs are expected. A good commercial chicken or turkey breeder mash will give satisfactory results, with birds either on range or in confinement, when fed in accordance with the manufacturer's directions. Clean, fresh water should always be available.

Egg production

The number of eggs a guinea hen will lay depends on her breeding and management. A hen that is of good stock and is carefully managed may lay 100 or more eggs a year.

Breeders generally produce well for two to three years; sometimes they are kept as long as four to five years in small farm flocks. In such flocks, hens will usually lay about 30 eggs and then go broody. If broken of broodiness, they may continue laying into the fall, and may produce from 50 to 100 eggs a year. Selection of breeders for egg and meat production traits, as practiced with chickens, would likely result in considerable improvement.

Guinea hens usually start laying in March or April, and may continue to lay until October. Then hens will lay in the house or in the yard if they are kept confined.

Breeders sometimes are kept confined during the day until about 3 p.m. or until they have laid. If allowed free range, they will make nesting places among the weeds and brush along the fences or in the fields. Two or three marked eggs should be left in these hidden or "stolen" nests so that the hens will continue to lay there. Gather the eggs daily, but do not disturb the hens while they are laying.

Keep the eggs in a cool place at 55° F., and do not hold them longer than one or two weeks before setting.

Guinea eggs are smaller than chicken eggs. They weigh about 1.4 ounces each; chicken eggs average about two ounces.

Incubation of eggs

The incubation period of guinea eggs is from 26 to 28 days; the incubation method is the same as for turkey eggs. Natural methods of incubation are generally used in small flocks; for large flocks, incubators are more satisfactory.

Chicken hens are more commonly used for hatching guinea eggs as they are more adaptable than guinea hens. Guinea hens usually are too wild to be set anywhere except in the nests where they have become broody.

As soon as some of the guinea keets hatch and begin moving about, the guinea hen is likely to leave the nest, abandoning the eggs that are not hatched. These eggs may yet hatch if, while still warm, they are put under another broody hen or in an incubator.

From 12 to 15 eggs may be set under a guinea hen; 20 to 28 may be set under a large chicken hen. Hens should be treated for lice before they are set.

Protect outside nests from the weather and from prowling animals. An individual run may be provided for each hen, or the hens may be taken off the nests daily and given feed and water.

Incubators

Two types of incubators may be used to incubate guinea eggs. One of these is known as a forced-draft incubator, in which a fan circulates the air. The other type has no fan and is known as a still-air machine.

The current temperature and humidity of the air within forced-draft incubators are about the same for both guinea and turkey eggs. Forced-draft incubators usually operate at about 99.5° to 99.7° F. and 57 to 58 percent humidity for the first three weeks.

Rearing keets

Chicken hens make the best mothers for guinea keets. Guinea hens are likely to take their keets through wet grass and lead them too far from home. Often, guinea hens will remain out in the fields at night instead of returning to their brood coops.

When two or more hens are set at one time, the keets may be doubled up at hatching time, and many keets hatched by guinea hens may be given to chicken hens to raise. A large chicken hen will brood as many as 25 guinea keets.

Place clean, dry litter in the coop. For the first two or three days the hen and keets should be confined to the coop, which may or may not have a small covered yard attached. After that time, the hen usually is confined and the keets are allowed to range, or both hen and keets may be allowed their freedom.

Brood coops should be closed at night to keep out predatory animals. The young keets should be kept confined in the morning until the grass dries. Allow the keets to range on clean grassland, and move the coops weekly or more often to fresh ground.

Guineas raised by natural methods usually will leave their coop when they are from six to eight weeks old, and will begin roosting at night in a nearby tree or other roosting place. They prefer roosting in the open, but if they have been raised with a chicken hen, they can be trained to follow her inside a poultry house to roost. If they have become accustomed to going into a house or other enclosure, they will not be so difficult to catch when they are wanted for the market.

Guineas often will remain close to the mother hen until they are almost fully grown. This attachment tends to control the natural wild instincts of the guineas and simplifies their production and management.

Brooder houses

Guinea keets are raised successfully in confinement in brooder houses that have wire-floored sun porches attached and equipment similar to that commonly used for raising young turkeys. They may be kept in these houses until they are ready for market. Flocks of as many as 200 keets are kept in brooder houses, and sometimes much larger numbers are raised in one group.

Most family flock guineas are raised 30 or 40 to a brooder. Careful sanitation and clean ground are important where good sized flocks of young guineas are raised.

Feeding

Guineas are fed much the same as turkeys. Their first feed may be turkey starting mash or crushed pellets upon which is scattered a little oatmeal or tender, finely chopped green feed. The starting mash should contain about 25 percent protein. Clean water always should be available.

Growing mash and grain may be fed after the keets are about six weeks old. During the first 10 days, either keep mash before the keets all the time or feed them four or five times daily. Usually, mash is kept before birds in confinement.

Young guineas will grow faster and be ready for market earlier if they are fed freely. Only two feeds a day need be given keets on range after they are well started. It is advisable to feed all flocks in late afternoon so they will return to their coops at night.

Marketing plans

A few small guineas usually start coming on the market late in June, and the general farm supply begins late in August. Thus, the normal marketing season is during the latter part of the summer and through the fall.

About half of the guinea fowl raised are for special and gourmet markets. Guinea fowl for special markets are sold alive and are primarily purchased by people of Asian descent. The gourmet birds are sold dressed and frozen to hotels and restaurants.

Frozen guinea fowl are almost always at least 15 weeks old and usually 16 to 18 weeks of age when they are sold. At this age their live weight is 2 ¾ to 3 ¼ pounds.

Preparing guineas for market

Most guineas are sold alive by the farmers to poultry processors. Then, the birds usually are dressed and scalded in the same way as chickens, except in very special cases, when they are marketed like game birds with the feathers left on. For all retail markets, as well as for hotel and restaurant trade, the feathers should be removed.

Guineas prepared for market by producers may be either scalded or dry picked. In dry picking, the roof of the mouth is severed first to insure thorough bleeding, and the knife is then thrust through the groove in the roof of the mouth into the brain. When the brain is pierced, the feathers are loosened by a convulsive movement of the muscles; this makes them easier to pick. If guineas are to be marketed with the feathers left on, all that needs to be done is to bleed the birds properly.

What type of poultry for you?

Chickens and bantams probably are the best choice for the home flock kept in a suburban area. They can be kept in confinement and many of the potential problems with neighbors can be avoided.

Ducks, geese, and turkeys are much bigger birds. They do not do as well in confinement as when out on range. They tend to create more mess than chickens and this can become a problem where neighbors are close.

Guinea fowl are very noisy. They set up loud squawks whenever a stranger comes around. They are not good neighbors. For this reason the birds are better kept on a small farm.

Chickens are the most available of the types of poultry for the home poultry flock. They are easiest to raise and eggs and fryers a ready market. The other birds can be raised by persons with average skill if you like duck, goose, or turkey meat. They are more of a challenge but the reward at the end may be worth the additional care.

Sources

THE EDITOR EXPRESSES APPRECIATION TO the U.S. Department of Agriculture and to many state universities for material used in this book. Poultry information has been excerpted and quoted from the following sources:

Abbott, Ursula K., Ralph A. Ernst, and Francine A. Bradley. *Incubating Eggs in Small Quantities.* Cooperative Extension, University of California, 2000.

Arrington, Louis C. *Should I Raise a Small Poultry Flock?* Cooperative Extension, University of Wisconsin, 1974.

Ash, William J. *Raising Ducks.* U.S. Department of Agriculture, Agricultural Research Service, 1971.

Ernst, Ralph A. and Gary Beall. *Why My Hens Stopped Laying.* Cooperative Extension, University of California, 1978.

Ernst. Ralph A., Pran Vohra, and Gary Beall. *Feeding Chickens.* Cooperative Extension, University of California, 1983.

Ernst, Ralph A., W. Stanley Coates, and Roderick A. Shipley. *Starting and Managing Small Poultry Units.* Cooperative Extension, University of California, 1974.

Geiger, Glenn and Harold Biellier. *Brooding and Rearing Ducklings and Goslings.* Cooperative Extension, University of Missouri, 1993.

Geiger, Glenn and Melvin L. Hamre. *Raising Geese.* U.S. Department of Agriculture, Agricultural Research Service, 1972.

Hamre, Melvin L. *The Small Laying Flock, Raising Geese.* Cooperative Extension, University of Minnesota, 1998.

Holleman, Kenneth A. *Managing the Pullet Flock.* Cooperative Extension, Oregon State University, reprinted 2005.

Jordan, Herbert C. *Home Processing of Poultry.* College of Agriculture Extension Service, Pennsylvania State University.

McClune, E. L. and Joseph M. Vandepopuliere. *Control of Poultry Disease Outbreaks.* Cooperative Extension, University of Missouri, 1993.

Mills, W. C. Jr., Thomas D. Siopes, and Thomas A. Carter. *Lighting System for Layers.* The North Carolina Agricultural Extension Service, North Carolina State University, 1981.

Newell, George, Pat Lewis, and A. L. Malle. *Poultry for the Small Producer.* Division of Agricultural Sciences and Natural Resources, Oklahoma State University.

Purchase, Graham H. *Farm Poultry Management.* U.S. Department of Agriculture, Agricultural Research Service, 1977.

Ridlen, S. F. and H. S. Johnson. *From Egg to Chick.* Cooperative Extension, University of Illinois, 1964.

Ringrose, Arthur T., and Denver D. Bragg. *Feather Picking and Cannibalism.* Agricultural Extension Service, Virginia Polytechnic Institute, 1969.

Rooney, W. F., Stanley Coates, and J. Price Schroeder. *Growing a Small Flock of Turkeys.* Volume 2733 of Leaflet Division of Agricultural Sciences, University of California, 1975.

Sheppard. C. C., Cal J. Flegal, and Thomas Thorburn. *The Small Poultry Flock.* Cooperative Extension, Michigan State University, 1974.

Skinner, John L. and Douglas A. Yanggen. *Raising Small Animals and Fowl in Urban Areas.* Cooperative Extension, University of Wisconsin, 1974.

Skinner, John L. *Raising Poultry for Eggs, Chicken Breeds and Varieties.* North Central Regional Extension Publications.

Talmadge, Daniel W. *Brooding and Rearing Baby Chicks.* Cooperative Extension Services of the Northeastern States, 1974.

Thacker, G. H. *The Home Poultry Flock.* A Cornell Cooperative Extensive Publication, reprinted 1996.

Index

Italicized page numbers refer to charts or illustrative material.

C

ALSO AVAILABLE

The Illustrated Guide to Chickens

How to Choose Them, How to Keep Them

by Celia Lewis
Introduction by HRH Prince Charles

Chickens are fun, useful, and easy to keep. If you have ever considered raising your own backyard flock, *The Illustrated Guide to Chickens* is the book for you! It offers practical advice and contains all the information you need to choose from one of the one hundred most familiar breeds of chicken in North America and Europe to raise. Each breed's profile is written in engaging text that covers its history and main characteristics. You'll also find practical advice about poultry rearing and husbandry and the pros and cons of pure breeds, hybrids, bantams, game foul, and the like.

$16.95 Hardcover • ISBN 978-1-61608-425-7

ALSO AVAILABLE

The Joy of Keeping Chickens

The Ultimate Guide to Raising Poultry for Fun or Profit

by Jennifer Megyesi

Photography by Geoff Hansen

Finally backyard farmers who want to keep a few hens for eggs have a bible that's attractive enough to leave out on the coffee table, and inexpensive enough to purchase on a whim. This comprehensive guide, written in charming prose from the perspective of an organic farmer, will appeal to readers who are interested in raising chickens, or simply want the best knowledge about how to cook them. With this in mind, farmer and animal expert Jennifer Megyesi discusses all the basic details of raising the birds—general biology, health, food, choosing breeds, and so on—and she cuts through the smoke to identify what terms like "organic," "free-range," and so on really mean for poultry farmers and consumers.

No chicken book would be complete without information on how to show chickens for prizes, and this is no different, but *The Joy of Keeping Chickens* also stresses the importance of self-sustainability and organic living, and the satisfaction of keeping heirloom breeds. Coupled with Geoff Hansen's gorgeous full-color photographs, this text makes for an instant classic in the category.

$14.95 Paperback • ISBN 978-1-60239-313-4

ALSO AVAILABLE

The Joy of Keeping Farm Animals

Raising Chickens, Goats, Pigs, Sheep, and Cows

by Laura Childs

When the going gets rough, the rough . . . start raising their own food. In the first full-color guide of its kind, author and small farm owner Laura Childs reveals exactly what it takes to start raising your own animals, including chickens, geese, goats, sheep, pigs, and cows. Childs discusses what you can expect to harvest from your animals—from eggs to milk to meat to wool—based on her own real-life experiences. Whether you want to raise a few chickens for eggs alone, try your hand at a few goats with the aim to make your own cheese, or are looking to sustain your family and make some extra money from raising and selling beef, this is the book for you.

Childs offers general information for each breed and animal, from how to get started to what to feed and where to house the animals. This invaluable guide is the perfect first book for anyone interested in starting a backyard barnyard or a small farm—or simply dreaming about the idea.

$14.95 Paperback • ISBN 978-1-60239-745-3

ALSO AVAILABLE

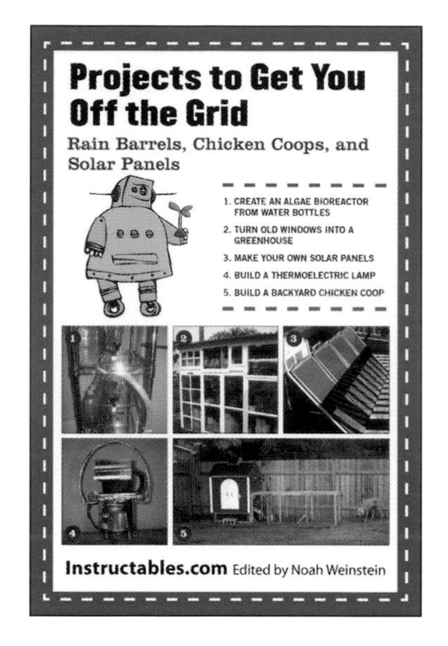

Projects to Get You Off the Grid

Rain Barrels, Chicken Coops, and Solar Panels

Edited by Noah Weinstein

Instructables is back with this compact book focused on a series of projects designed to get you thinking creatively about thinking green. Twenty Instructables illustrate just how simple it can be to make your own backyard chicken coop, or turn a wine barrel into a rainwater collector.

Illustrated with dozens of full-color photographs per project accompanying easy-to-follow instructions, this Instructables collection utilizes the best that the online community has to offer, turning a far-reaching group of people into a mammoth database churning out ideas to make life better, easier, and in this case, greener, as this volume exemplifies.

NOTES

NOTES

NOTES